NUREG-1903

Seismic Considerations For the Transition Break Size

Manuscript Completed: February 2008
Date Published: February 2008

Prepared by
N.C. Chokshi[1], S.K. Shaukat[1], A.L. Hiser[1], G. DeGrassi[2]
G. Wilkowski[3], R. Olson[4], J.J. Johnson[5]

[2]Brookhaven National Laboratory
Building 130
P.O. Box 5000
Upton, NY 11973

[3]Engineering Mechanics Corporation of Columbus
3518 Riverside Drive, Suite 202
Columbus, OH 43221-1735

[4]Battelle
505 King Avenue
Columbus, OH 43201-2693

[5]James J. Johnson & Associates
7 Essex Court
Alamo, CA 94507

[1]Office of Nuclear Regulatory Research

ABSTRACT

The U.S. Nuclear Regulatory Commission (NRC) has been considering revision of the regulatory requirements for the emergency core cooling system (ECCS), as set forth in Title 10, Section 50.46, of the *Code of Federal Regulations* (10 CFR 50.46); Appendix K to 10 CFR Part 50; and General Design Criterion (GDC) 35. In particular, those requirements state that the ECCS shall be sized to provide adequate makeup water to compensate for a break of the largest diameter pipe in the primary system [i.e., the so-called "double-ended guillotine break" (DEGB)]. Consequently, in order to risk-inform the break size, the staff of the NRC proposed the concept of transition break size (TBS). In addition, the NRC developed pipe break frequencies as a function of break size using an expert elicitation process for degradation-related pipe breaks in reactor coolant systems of typical boiling- and pressurized-water reactors. That elicitation focused on determining event frequencies that initiate by unisolable primary side failures that can be exacerbated by material degradation with age under normal operating conditions. The purpose of this study was to assess potential seismic effects on the postulated TBS, and to provide information to facilitate review and comment regarding the proposed risk-informed revision of ECCS requirements in 10 CFR 50.46. Thus, this report evaluates the seismic effects, using different approaches to evaluate flawed and unflawed piping, and indirect failures of other components and component supports that could lead to piping failure.

FOREWORD

U.S. Nuclear Regulatory Commission (NRC) requirements state, in part, that the emergency core cooling system (ECCS) shall be sized to provide adequate make-up water to compensate for a break of the largest diameter pipe in the primary system [i.e., the so-called "double-ended guillotine break" (DEGB)]. However, the DEGB is recognized as an extremely unlikely event. The NRC is working to finalize a rule change that would allow for a risk informed alternative to the present DEGB.

To provide the technical basis for a risk-informed alternative to the current regulatory requirements, and for use in a variety of regulatory applications, including probabilistic risk assessments (PRAs), the NRC staff developed loss-of-coolant accident (LOCA) frequency estimates using an expert elicitation process. This process consolidated service history data and insights from probabilistic fracture mechanics studies with knowledge of plant design, operation, and material performance to estimate break frequencies. Expert elicitation is a well-recognized process for quantifying phenomenological knowledge when modeling approaches or data are insufficient.

On the basis of the expert elicitation, the NRC staff established a baseline break size corresponding to a break frequency of once per 100,000 years (i.e., 10^{-5} per year). The staff then adjusted this baseline break size to account for other significant contributing factors that were not explicitly addressed in the expert elicitation to define an alternative risk-informed break size, termed the transition break size (TBS). In addition, because the elicitation did not include effects of rare event loadings, such as seismic events, a separate study was undertaken to assess potential seismic effects on the postulated TBS. The results of this study are summarized in this report.

To evaluate seismic effects, the staff used different techniques to evaluate flawed and unflawed piping, and indirect failures of other components and component supports that could lead to piping failure. For flawed and unflawed piping, the approach used is a hybrid of deterministic and probabilistic analysis. In particular, for flawed piping the study determined the maximum flaw size in large piping systems that would not fail under stresses resulting from seismic loads with a frequency of occurrence of 10^{-5} per year and 10^{-6} per year. These results provide insight into the extent of degradation that would have to be present in the reactor coolant system piping to affect the TBS.

This study demonstrated that the critical flaws associated with the stresses induced by seismic events with an annual probability of exceedance of 10^{-5} and 10^{-6} are generally large, and coupled with other mitigating aspects, the probabilities of pipe breaks larger than the TBS are likely to be less than 10^{-5} per year. Since it is extremely difficult to model the time dependent nature of failure probabilities of passive components that are subject to time-dependent degradation and accidental loads in a PRA, the approach used in this study can serve as an example of approach that could be used in other efforts to determine the extent to which degradation specific to a component under consideration becomes risk-significant. However the results of this study should not be applied beyond components addressed in this report.

In summary, this study demonstrated that although the probability of seismically induced failures of reactor coolant piping system may vary by site, selection of a TBS is unlikely to be affected by seismic considerations. This study was made publicly available on NRC's web site on

December 15, 2005, in conjunction with the proposed rule to facilitate public review and comment. This report is now being published to make it more widely available.

CONTENTS

ABSTRACT iii
FOREWORD v
CONTENTS vii
EXECUTIVE SUMMARY xi
ACKNOWLEDGEMENTS xv
ABBREVIATIONS xvii
1 INTRODUCTION 1-1
   1.1 Transition Break Size: Definition and Basis 1-1
   1.2 Summary of Expert Elicitation and Resulting Transition Break Size 1-2
   1.3 Objective and Approach 1-4
2 BACKGROUND AND OVERVIEW OF STAFF APPROACH 2-1
   2.1 Background 2-1
   2.2 Overview of the Staff Approach 2-2
   2.3 Review of Past Seismic Experience Data 2-4
   2.4 Review of Probabilistic Risk Assessments (PRAs) 2-5
3 SUMMARY OF THE LLNL STUDY 3-1
   3.1 Objective 3-1
   3.2 Overall Scope and Approach 3-2
      3.2.1 Approach: Direct DEGB 3-4
      3.2.2 Approach: Indirect DEGB 3-7
   3.3 Key Findings 3-7
      3.3.1 Findings: Direct DEGB 3-7
      3.3.2 Findings: Indirect DEGB 3-8
   3.4 Relationship of the LLNL Study to the Present Effort 3-9
4 THE STAFF'S APPROACH FOR THE PRESENT STUDY 4-1
   4.1 General Approach and Scope of Analyses 4-1
   4.2 Description of the LBB Database 4-3
   4.3 Seismic Hazard 4-5
   4.4 Unflawed Piping 4-7
      4.4.1 Analysis – Approach and Key Steps 4-7
      4.4.2 Description of the Scale Factor Approach 4-10
      4.4.3 Brief Description of Unflawed Piping Failure Criterion 4-21
      4.4.4 Key Results and Findings 4-22
   4.5 Flawed Piping 4-23
      4.5.1 Approach, Assumptions, and Key Steps 4-23
      4.5.2 Surface Flaw Evaluation Procedures 4-26
      4.5.3 Through-Wall Flaw Evaluation Procedures for LBB Analysis 4-41
      4.5.4 Overall Results of Flawed-Piping Analysis 4-48
   4.6 Indirectly Induced Piping Failures 4-48
      4.6.1 Overview and Background 4-48
      4.6.2 CE Plant 4-49
      4.6.3 Westinghouse Plant 4-49
      4.6.4 Observations 4-50
5 SUMMARY AND CONCLUSIONS 5-1
   5.1 Summary 5-1
   5.2 Conclusions and Overall Assessment 5-3
6 REFERENCES 6-1

Figures

Figure 1-1 Estimates of primary system pipe break frequency 1-3
Figure 1-2 Schematic of seismic-induced break frequencies 1-4
Figure 3-1 Generic seismic hazard curves for plants east of the Rocky Mountains 3-3
Figure 3-2 Flow chart for probabilistic assessment of piping integrity 3-5
Figure 4-1 Seismic hazard curves for 43 PWR plant sites east of the Rocky Mountains 4-6
Figure 4-2 Approach and key steps for unflawed pipe evaluation 4-7
Figure 4-3 Probability of exceedance versus maximum normal + seismic stress in RCS piping at a hypothetical PWR 4-9
Figure 4-4 Probability of exceedance versus maximum normal + seismic stress in RCS piping at 27 PWRs 4-22
Figure 4-5 Probability of exceedance versus maximum normal + seismic stress in pressurizer surge piping at six PWRs 4-23
Figure 4-6 Critical Flaw Size Determination Procedure 4-25
Figure 4-7 Simple elastic-stress correction curves for typical nuclear piping steels using Code strength values (Note that a lower correction would exist using actual yield strength values, which are typically higher than Code values.) 4-29
Figure 4-8 Comparison of ASME Code and Best-Estimate Z-factors for ferritic steels in Plant C cases 4-34
Figure 4-9 Comparison of ASME Code and Best-Estimate Z-factors for austenitic base metals and welds in Plant A and B cases and the Plant C cold-leg case 4-34
Figure 4-10 Comparison of ASME maximum allowable flaw sizes to *Best-Estimate* critical flaw sizes at 10^{-5} annual probability of exceedance seismic event for Plant A hot-leg case 4-36
Figure 4-11 Example of Category A results for a hot leg 4-38
Figure 4-12 Example of Category B results for a crossover leg at the 10^{-5} annual probability of exceedance seismic event 4-38
Figure 4-13 Example of Category C results for cold leg on the discharge side of a CE plant at the 10^{-5} annual probability of exceedance seismic event 4-39
Figure 4-14 Critical surface flaw a/t values versus N+10^{-5} stresses for θ/π = 0.8 (For all loops and plants and all materials together) 4-40
Figure 4-15 Critical surface flaw a/t values versus N+10^{-6} per year seismic stresses for θ/π = 0.8 (For all loops and plants and all materials together) 4-40
Figure 4-16 Critical circumferential through-wall flaw length corresponding to 10^{-5} per year seismic stresses 4-41
Figure 4-17 Critical circumferential through-wall flaw length corresponding to 10^{-6} per year seismic stresses 4-44
Figure 4-18 Sensitivity study results showing leakage to critical crack size ratio versus normal to (N + 10^{-5} seismic) stress ratio for five different plant piping systems with different cracking mechanisms 4-46
Figure 4-19 Sensitivity study results showing leakage to critical crack size ratio versus normal to (N + 10^{-5} seismic) stress ratio for five different plant piping systems with safety factor of 1.5 on crack length and safety factor of 10 on leak rate 4-47
Figure 4-20 Sensitivity study results showing leakage to critical crack size ratio versus normal to (N + 10^{-6} seismic) stress ratio for five different plant piping systems with safety factor of 1.0 on crack length and either the 1 gpm technical specification leak rate or 0.5 gpm leak rate (both with a safety factor of 10) 4-47

Tables

Table 3-1	Annual Probabilities of Indirect DEGB for Events Per Year	3-8
Table 4-1	Information from the LBB Database	4-7
Table 4-2	Estimates of Normalized Stress Ratios and Probability of Exceedance	4-8
Table 4-3	Important Parameters in Seismic Methodology	4-15
Table 4-4	Factors of Safety: Structure Response for CE Plant	4-18
Table 4-5	Factors of Safety: Structure Response for Westinghouse Plant — Soil Site	4-19
Table 4-6	Factors of Safety: Structure Response for Westinghouse Plant — Rock Site	4-20
Table 4-7	Stress Values at 1% Failure Probability	4-21
Table 4-8	Multipliers on Monotonic-Loaded Quasi-Static J-R Curves for Dynamic and Cyclic Loading Effects for Seismic Applications	4-32
Table 4-9	Factors of Safety: RCL Major Component Response	4-49

x

EXECUTIVE SUMMARY

This report describes the results of a study, sponsored by the U.S. Nuclear Regulatory Commission (NRC), to assess potential seismic effects on the postulated transition break size (TBS) in the proposed risk-informed revision of the regulatory requirements for the emergency core cooling system (ECCS), contained in Title 10, Section 50.46, of the *Code of Federal Regulations* (10 CFR 50.46). This study was conducted by the staff of the NRC's Office of Nuclear Regulatory Research (RES), in conjunction with the NRC's Office of Nuclear Reactor Regulation (NRR), with technical support from Brookhaven National Laboratory (BNL), Engineering Mechanics Corporation of Columbus (Emc^2), Battelle Columbus, and James J. Johnson & Associates. The primary focus of this report is to provide sufficiently detailed discussion of the study's approach and results to facilitate review and comment concerning the proposed rule and statement of considerations (SOC), entitled "Risk-Informed Changes to Loss-of-Coolant Accident Technical Requirements; Proposed Rule," which the NRC published in the *Federal Register* (70 FR 67598) on November 7, 2005.

The regulatory requirements for the design, construction, and operation of the ECCS for light-water nuclear reactors are contained in 10 CFR 50.46, "Acceptance criteria for emergency core cooling systems for light-water nuclear reactors"; Appendix K to 10 CFR Part 50, "ECCS Evaluation Models"; and Appendix A to 10 CFR Part 50, "General Design Criteria [GDC] for Nuclear Power Plants" (e.g., GDC 35, "Emergency Core Cooling"). These requirements state, in part, that the ECCS shall be sized to provide adequate make-up water to compensate for a break of the largest diameter pipe in the primary system [i.e., the so-called "double-ended guillotine break" (DEGB)]. However, the DEGB is recognized as an extremely unlikely event. In order to risk-inform the break size, the staff proposed the concept of the TBS, as described in SECY-05-0052, "Proposed Rulemaking for 'Risk-Informed Changes to Loss-of-Coolant Accident [LOCA] Technical Requirements'," dated March 29, 2005, which included the following statement:

> *The proposed rule would divide the current spectrum of LOCA break sizes into two regions. The division between the two regions is determined by a "transition break size" (TBS). The first region includes small breaks up to and including the TBS. The second region includes breaks larger than the TBS up to and including the double-ended guillotine break (DEGB) of the largest reactor coolant system pipe. The term, "break," in the TBS does not mean a double-ended guillotine break; rather it refers to an equivalent opening in the reactor coolant system boundary.*

As discussed in the SOC, to help establish the TBS, the NRC developed pipe break frequencies as a function of break size using an expert elicitation process for degradation-related pipe breaks in reactor coolant systems (RCSs) of typical boiling- and pressurized-water reactors (BWRs and PWRs, respectively). This elicitation process is used to quantify phenomenological knowledge when data or modeling approaches are insufficient. The elicitation focused solely on determining event frequencies that initiate by unisolable primary system side failures related to material degradation. Results of this elicitation are discussed in NUREG-1829, "Estimating Loss-of-Coolant Accident (LOCA) Frequencies through the Elicitation Process," which the NRC published in June 2005 [NRC, 2005].

On the basis of the expert elicitation, the NRC staff established a baseline TBS using these pipe break frequencies as a starting point. The staff then adjusted this baseline TBS to account for other significant contributing factors that were not explicitly addressed in the expert elicitation.

In the present study to evaluate seismic effects, the staff used different approaches in evaluating flawed and unflawed piping, and indirect failures of other components and component supports that could lead to piping failure. For flawed and unflawed piping, the approach used is a hybrid deterministic and probabilistic analysis, which addresses uncertainties (in part) through sensitivity analyses, rather than explicit uncertainty analysis. For indirect failures, the approach is based on an earlier study by Lawrence Livermore National Laboratory (LLNL) [NRC, 1985a, 1985b, 1985c, and 1988a].

Flawed Piping: The following are some of the higher-level observations, conclusions, and insights derived from results discussed in Section 4.5:

(1) The absolute size of the best-estimate critical surface flaw sizes are large for ground motion levels corresponding to seismic events associated with 10^{-5} and 10^{-6} annual probabilities of exceedance. (The present study used an earthquake frequency of 10^{-5} per year as a general point of comparison to the TBS, because the frequency of 10^{-5} per year was a basis for establishing the TBS.) For the cases analyzed, even for very long circumferential cracks, the surface flaw depth must be larger than 40% of the wall thickness in order to become critical for stresses associated with a seismic event with a 10^{-5} annual probability of exceedance. The corresponding surface flaw depth is 30% of the wall thickness for a seismic event with a 10^{-6} annual probability of exceedance. This is the most significant finding, in that these surface flaw sizes are large enough that either an effective inspection program or a leak detection system can be implemented to ensure that a surface flaw will be detected in time and will not grow to critical size during service.

(2) Critical flaw sizes depend on applied stresses and material strength parameters and vary significantly for different locations and piping systems. Both normal and seismic stresses depend upon layout configuration, design process, support designs, site seismic hazard and ground motion characteristics, and other plant-specific features.

(3) The flaw sizes allowed by the Boiler and Pressure Vessel (BPV) Code promulgated by the American Society of Mechanical Engineers (ASME) (see footnote *** in Section 5.1) are affected by whether one uses ASME Section II material properties or actual material properties based on availability of test data.

(4) Comparisons of best-estimate versus Code-allowable flaw size curves for analyzed cases fall into three categories: (a) best-estimate flaw sizes are bigger than the maximum flaw sizes allowed by the Code using either Code material properties or actual material properties; (b) best-estimate flaw sizes fall between two Code flaw size estimates; and (c) best-estimate critical flaw sizes are smaller than both Code estimates. Thus, the best-estimate flaw sizes are not necessarily larger than the Code flaw sizes in all cases; however, more frequently, the maximum flaw size by the Code was smaller than the best-estimate critical flaw sizes.

(5) Piping systems of BWR and west coast plants have not been studied in this report because required information was not readily available. However, there are no inherent limitations in applying the approach used here to piping systems of BWR and west coast plants.

Unflawed Piping: Analyses performed in this study show that seismic-induced failure probabilities of unflawed piping, as defined in Section 2, are significantly low compared to the frequency of 10^{-5} per year used as a basis to establish the TBS.

Indirect Failures: The present study yielded the following key observations and conclusions:

(1) As with other parts of the seismic analysis, indirect failure evaluations are plant- and site-specific. Details of plant layout and component and pipe support designs vary significantly from plant to plant.

(2) For the two cases analyzed, indirectly induced piping failure attributable to major component support failure has a mean failure probability on the order of 10^{-6} per year.

In summary, this study demonstrated that the critical flaws associated with the stresses induced by seismic events with an annual probability of exceedance of 10^{-5} and 10^{-6} are large and, coupled with other mitigative aspects, the probabilities of pipe breaks larger than the TBS are likely to be less than 10^{-5} per year. Similarly, for the cases studied, the probabilities of indirect failures of large RCS piping systems are less than 10^{-5} per year.

The intent of the staff study was not to perform bounding analyses that will encompass all potential variations, including site-to-site variability in the seismic hazard. The purpose of the staff study was to obtain a measure of seismic effects on the proposed TBS and to provide information on key considerations to facilitate the public review and comment period to elicit comments and information germane to this issue.

ACKNOWLEDGEMENTS

The authors would like to thank the following members of the NRC staff, who provided assistance regarding the technical issues and participated in the development of this report:

- Mr. Charles (Gary) Hammer
- Mr. Adam Wilson
- Mr. John Fair
- Mr. C.E. (Gene) Carpenter
- Dr. Mark T. Kirk

The authors also acknowledge and appreciate the efforts of Dr. Mano Subudhi of Brookhaven National Laboratory, Messrs. Brett Burton and Paul Scott of Battelle, and Drs. Heqin Xu and Prabhat Krishnaswamy of Engineering Mechanics Corporation of Columbus.

ABBREVIATIONS

ASME	American Society of Mechanical Engineers
B&W	Babcock & Wilcox
BNL	Brookhaven National Laboratory
BPV	Boiler and Pressure Vessel Code (ASME)
BWR	boiling-water reactor
CE	Combustion Engineering
CMTR	Certified Material Test Report
COD	crack-opening displacement
DEGB	double-ended guillotine break
DGRS	design ground response spectra
DOE	U.S. Department of Energy
ECCS	emergency core cooling system
EERI	Earthquake Engineering Research Institute
Emc^2	Engineering Mechanics Corporation of Columbus
EPRI	Electric Power Research Institute
IGSCC	intergranular stress-corrosion cracking
IPEEE	Individual Plant Examination of External Events
IPIRG	International Piping Integrity Research Group
ISI	inservice inspection
LBB	leak-before-break
LLNL	Lawrence Livermore National Laboratory
LOCA	loss-of-coolant accident
LWR	light-water reactor
NIST	National Institute of Standards and Technology
NRC	U.S. Nuclear Regulatory Commission
NSSS	nuclear steam supply system
PGA	peak ground acceleration
PRA	probabilistic risk assessment
PWR	pressurized-water reactor
PWSCC	primary water stress-corrosion cracking
RCL	reactor coolant loop
RCS	reactor coolant system
RHR	residual heat removal
SAM	seismic anchor motion
SAW	submerged arc weld
SF	Scale Factor
SMA	seismic margin assessment
SMAW	shielded metal arc weld
SOC	statement of considerations
SQUIRT	Seepage Quantification in Reactor Tubing
SRP	Standard Review Plan
SSE	safe shutdown earthquake
SSI	soil-structure interaction
SSMRP	Seismic Safety Margin Research Program
TBS	transition break size
TIG/MIG	tungsten inert gas/metal inert gas
UHS	uniform hazard spectra

1 INTRODUCTION

This report describes the results of a study, sponsored by the U.S. Nuclear Regulatory Commission (NRC), to assess potential seismic effects on the postulated transition break size (TBS) in the proposed risk-informed revision of the regulatory requirements for the emergency core cooling system (ECCS) contained in Title 10, Section 50.46, of the *Code of Federal Regulations* (10 CFR 50.46). This study was conducted by the staff of the NRC's Office of Nuclear Regulatory Research (RES), in conjunction with the NRC's Office of Nuclear Reactor Regulation (NRR), with technical support from Brookhaven National Laboratory (BNL), Engineering Mechanics Corporation of Columbus (Emc^2), Battelle Columbus, and James J. Johnson & Associates. The primary focus of this report is to provide sufficiently detailed discussion of the study's approach and results to facilitate review and comment concerning the proposed rule and statement of considerations (SOC), entitled "Risk-Informed Changes to Loss-of-Coolant Accident Technical Requirements; Proposed Rule," which the NRC published in the *Federal Register* (70 FR 67598) on November 7, 2005.

Sections 1.1 and 1.2 of this report summarize the pertinent portions of the proposed rule and SOC to provide essential background and context for the technical and regulatory issues discussed in this report. Section 1.3 describes the objective and approach used for this research.

1.1 Transition Break Size: Definition and Basis

The regulatory requirements for the design, construction, and operation of the ECCS for light-water nuclear reactors are contained in 10 CFR 50.46, "Acceptance criteria for emergency core cooling systems for light-water nuclear reactors"; Appendix K to 10 CFR Part 50, "ECCS Evaluation Models"; and Appendix A to 10 CFR Part 50, "General Design Criteria [GDC] for Nuclear Power Plants" (e.g., GDC 35, "Emergency Core Cooling"). These requirements state, in part, that the ECCS shall be sized to provide adequate make-up water to compensate for a break of the largest diameter pipe in the primary system [i.e., the so-called "double-ended guillotine break" (DEGB)]. However, the DEGB is recognized as an extremely unlikely event. In order to risk-inform the break size, the staff proposed the concept of the TBS, as described in SECY-05-0052, "Proposed Rulemaking for 'Risk-Informed Changes to Loss-of-Coolant Accident [LOCA] Technical Requirements'," dated March 29, 2005, which included the following statement:

> *The proposed rule would divide the current spectrum of LOCA break sizes into two regions. The division between the two regions is determined by a "transition break size" (TBS). The first region includes small breaks up to and including the TBS. The second region includes breaks larger than the TBS up to and including the double-ended guillotine break (DEGB) of the largest reactor coolant system pipe. The term, "break," in the TBS does not mean a double-ended guillotine break; rather it refers to an equivalent opening in the reactor coolant system boundary.*

Similarly, the proposed rule defines the TBS as follows:

> *Transition break size (TBS) is a break of area equal to the cross-sectional flow area of the inside diameter of specified piping for a specific reactor. The specified piping for a pressurized-water reactor is the largest piping attached to the reactor coolant system. The specified piping for a boiling-water reactor is the larger of the feedwater line inside containment or the residual heat removal line inside containment.*

As discussed in the SOC, to help establish the TBS, the NRC developed pipe break frequencies as a function of break size using an expert elicitation process for degradation-related pipe breaks in reactor coolant systems (RCSs) of typical boiling- and pressurized-water reactors (BWRs and PWRs, respectively). This elicitation process is used to quantify phenomenological knowledge when data or modeling approaches are insufficient. This elicitation focused solely on determining event frequencies that initiate by unisolable primary system side failures related to material degradation. Results of this elicitation are discussed in draft NUREG-1829, "Estimating Loss-of-Coolant Accident (LOCA) Frequencies through the Elicitation Process," which the NRC published in June 2005 [NRC, 2005].

On the basis of the expert elicitation, the NRC staff established a baseline TBS using these pipe break frequencies as a starting point. The staff then adjusted this baseline TBS to account for other significant contributing factors that were not explicitly addressed in the expert elicitation process. Thus, the staff used the following three-step process in establishing the TBS:

(1) Select break sizes for each light-water reactor type (i.e., BWR and PWR), corresponding to a break frequency of once per 100,000 years (i.e., 10^{-5} per year) from the expert elicitation results.

(2) Consider uncertainty in the elicitation process, other potential mechanisms that could cause pipe failure and were not explicitly considered in the expert elicitation process, and the higher susceptibility to rupture or failure of specific piping in the RCS.

(3) Adjust the TBS (upward) to allow necessary conservatism in order to account for the uncertainties.

1.2 Summary of Expert Elicitation and Resulting Transition Break Size

The aim of the elicitation described in draft NUREG-1829 [NRC, 2005] was to consolidate service history data and insights from probabilistic fracture mechanics studies with knowledge of plant design, operation, and material performance. The scope of the LOCA frequency estimates was restricted as follows:

- **Piping Location**
    - Addressed: The elicitation focused on determining event frequencies that are initiated by unisolable primary system side failures, as well as the potential exacerbating effects of age-related material degradation on those frequency estimates.
    - Not Addressed: The elicitation did not address LOCA frequency arising from active system failures (e.g., stuck open valve, pump seals, interfacing system LOCAs), and consequential primary pressure boundary failures attributable to either secondary side failures or failures of other plant structures (e.g., crane drops).

- **Loading Type**
    - Addressed: Consideration was limited to normal plant operational cycles and loading histories, as well as expected transient stresses that occur over the extended licensing period. However, the elicitation explicitly addressed only those loading events with frequencies greater than approximately 0.01/calender year.
    - Not Addressed: The elicitation did not consider loading caused by "rare" events (e.g., earthquake loading, water hammer, terrorist activity, etc.) because of the strong dependence on plant-specific factors.

Figure 1-1 illustrates the summary results of this investigation, which draft NUREG-1829 [NRC, 2005] presents as the variation in yearly pipe break frequency with piping diameter for BWRs and PWRs.

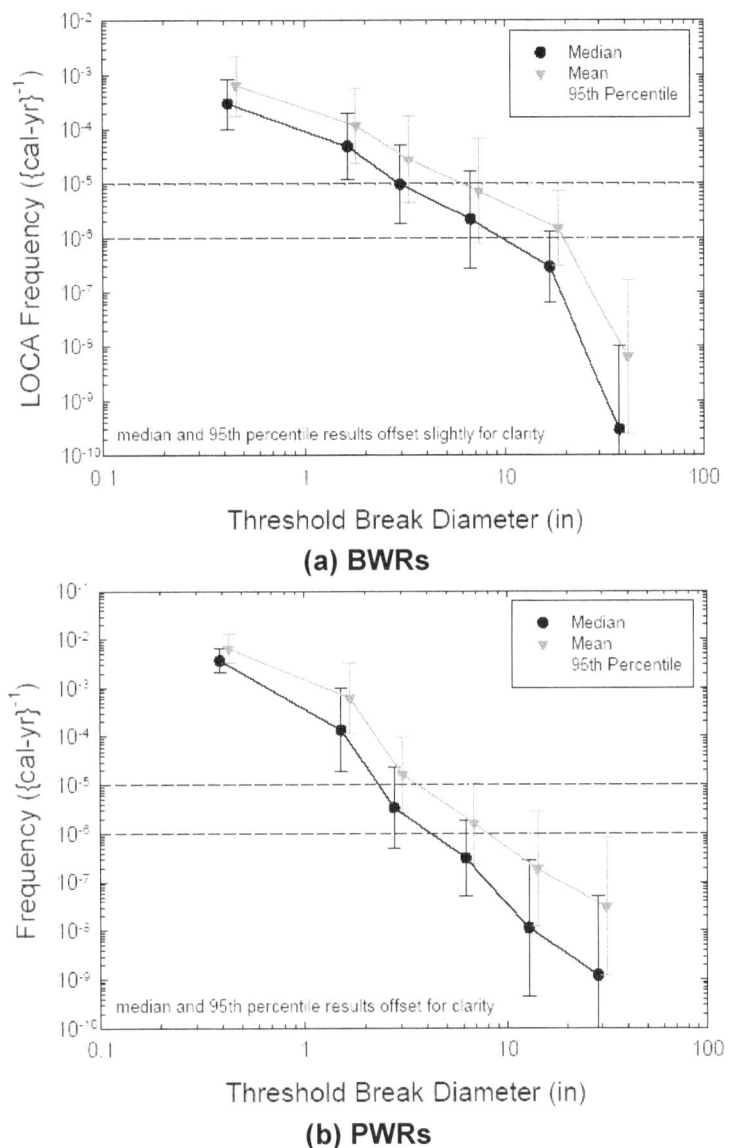

Figure 1-1   Estimates of Primary System Pipe Break Frequency

In addition to the above baseline estimates, draft NUREG-1829 [NRC, 2005] documents several areas of uncertainties associated with the expert elicitation process and the techniques used in combining individual expert input. (The report also includes the results of investigations to address these uncertainties.) Therefore, in defining the TBS, the staff began with a break frequency of 10^{-5} per year and adjusted upward in accordance with the three-step process discussed in Section 1.1. For PWRs, the adjusted TBS is a break with a diameter on the order of 12 to 14 inches (30.5 to 35.6 cm), controlled primarily by the surge line. By contrast, for BWRs, the adjusted TBS is a break with a diameter on the order of 20 inches (50.8 cm). It should be noted that the resulting TBSs for PWRs and BWRs do not directly relate to a specific break frequency, but are expected to have a frequency less than 10^{-5} per year.

1.3 Objective and Approach

Of all rare-event loadings listed in Section 1.2 and discussed in detail in the SOC, the only event having a credible potential to contribute significantly to the LOCA break frequency is earthquake loading. Therefore, the objective of this investigation (and the subject of this report) is to assess the adequacy of the proposed TBS when the effects of seismic loading are considered.

Ideally, it would be desirable to produce results in terms of the conditional probabilities of break sizes for piping of various diameters, given a certain ground motion level of a postulated seismic event, and then determine the absolute frequency of break sizes by coupling the conditional probabilities with the seismic hazard information (i.e., frequency of occurrence of seismic events of various ground motion levels for a given site). This is illustrated schematically in Figure 1-2 (without uncertainty distribution).

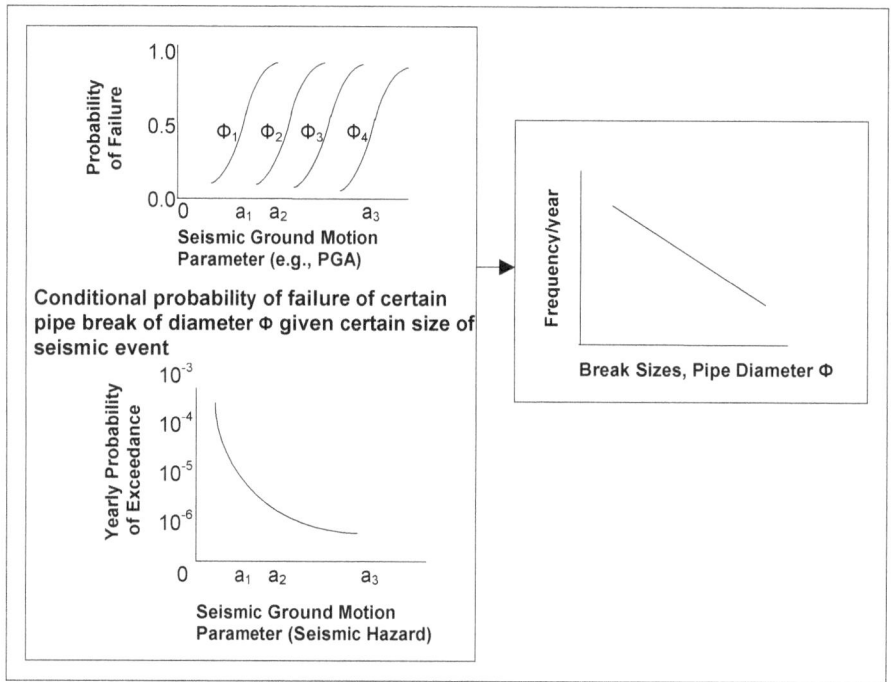

Figure 1-2 Schematic of Seismic-Induced Break Frequencies

This process would then allow direct comparison of the results from seismic-induced LOCA frequency of various break sizes with the results in draft NUREG-1829 [NRC, 2005]. However, such an effort would involve considerable time and resources. As noted in the SOC, seismic-induced frequencies are highly dependent on the site-specific hazard, plant design, and configurations. A generic study would have to encompass a wide range of variations applicable to the entire population of plants. Section 3 describes such a general approach, as used in a previous study.

In the present study, the staff adopted an alternative hybrid deterministic and probabilistic approach with the following two principal objectives:

(1) Examine the likelihood and conditions that would result in the prediction of seismic-induced breaks that are incompatible with the proposed TBS. (The present study used an earthquake frequency of 10^{-5} per year as a general point of comparison to the TBS, because the frequency of 10^{-5} per year was a basis for establishing the TBS.)

(2) Develop analytical procedures that can be used to perform case-specific analyses.

In addition to time and resources, the governing factors in adopting the aforementioned approach are as follows:

- highly site-specific and plant-specific frequencies of seismic-induced LOCAs
- ongoing and recent developments in seismic hazard studies (for example, as part of early site permit activities)
- availability of plant-specific information such as normal operating stresses, design seismic stresses, and material properties

As discussed later in this report, these factors influenced the scope of the study.

2 BACKGROUND AND OVERVIEW OF STAFF APPROACH

2.1 Background

The primary loop and connected piping systems in nuclear power plants are designed in accordance with the requirements set forth in Section III of the Boiler and Pressure Vessel (BPV) Code promulgated by the American Society of Mechanical Engineers (ASME). As part of the design, the seismic stresses induced by the safe shutdown earthquake (SSE) are combined with the normal operating stresses to ensure that the piping system can withstand these stresses with adequate safety margin. Inservice inspection (ISI) and the disposition of any flaws found during the inspections are done in accordance with the requirements set forth in Section XI of the ASME BPV Code.

The issue of seismic response of nuclear piping systems has been studied domestically and internationally over many years from various perspectives such as establishing design rules, developing and validating analytical models, assessing the behavior of flawed and unflawed piping*, leak-versus-break behavior, and failure modes under various types of loadings. These studies, which included testing, analysis, evaluation of piping system performance in earthquakes, and probabilistic risk assessment (PRA), include the following examples:

- "Seismic Analysis of Piping," NUREG/CR-5361 [Jaquay, 1998]
- "International Piping Integrity Research Group (IPIRG) Program, Final Report" NUREG/CR-6233, Vol. 4 [Wilkowski et al., 1997]
- "Review of Seismic Response Data for Piping" [Slagis, 1995]
- "Methodology for Developing Seismic Fragilities" [Reed and Kennedy, 1994]
- "Individual Plant Examination of External Events (IPEEE) Seismic Insights" [EPRI, 2000]
- "Survey of Strong Motion Earthquake Effects on Thermal Power Plants in California with Emphasis on Piping Systems," Main Report, Vol. 1, NUREG/CR-6239 [Stevenson, 1995]

One of the more significant and directly related studies, from the perspective of the present study, was the NRC-sponsored program in the 1980s at Lawrence Livermore National Laboratory (LLNL), which estimated the probability of an RCS DEGB for each of the four U.S. reactor vendor designs [i.e., Westinghouse, Combustion Engineering (CE), General Electric (GE), and Babcock & Wilcox (B&W)]. The objective of this program was to estimate the probability of DEGB attributable to (1) pipe fracture directly caused by the growth of cracks at welded joints, termed "direct DEGB," and (2) pipe rupture indirectly caused by failure of major components or component supports as a result of an earthquake, termed "indirect DEGB." These studies are documented in a variety of NUREG-series reports [NRC, 1985a, 1985b, 1985c, 1988a].

---

* For the purposes of this study, a piping system is considered "unflawed" when flaws are smaller than the maximum allowable workmanship flaws described in the ASME Section XI flaw standards in the IWB-3514 tables. Such flaws are so small that, under monotonic loading, the flawed pipe behaves as if its entire cross-section is effective in resisting applied loads (i.e., the pipe would buckle without the flaw growing). Failure of piping under seismic-type loading occurs primarily by fatigue-ratcheting as a crack initiates and grows during the application of cyclical loads. By contrast, "flawed" pipe has flaws large enough so that the entire nominal cross-section area is not effective in resisting applied loads, and pipes fail by either net-section collapse (e.g., the available cross-sectional area becoming fully plastic) or ductile tearing (e.g., flaw tearing with a low fracture toughness through the remaining thickness and possibly growing around the circumference).

In general and based in part on studies such as those noted above, the past seismic design of piping systems is thought to be conservative with ample margins. The actual earthquake experience of non-nuclear power plant piping, tests, and PRA studies also supports this position. Sections 2.3 and 2.4 of this report provide additional detail concerning observations from past earthquakes and PRA insights.

2.2 Overview of the Staff Approach

With the above background, the staff adopted an approach with two main objectives to (1) examine the likelihood and conditions that would result in seismic-induced breaks that are incompatible with the proposed TBS, and (2) develop analytical procedures that can be used to perform case-specific analyses. In the preliminary stages of this project, NRC staff developed the following list of technical issues to be addressed:

(1) Are seismic-induced *undegraded* passive component failure frequencies significant, compared to the expert elicitation findings for normal operating stresses?

(2) Are seismic-induced *degraded* passive component failure frequencies significant, compared to the expert elicitation findings for normal operating conditions?

(3) Does the significance of seismic-induced component failure frequencies increase or decrease with respect to the expert elicitation findings[†] as a function of LOCA size for undegraded and/or degraded piping?

(4) What are the most important generic and plant-specific considerations that affect the seismic-induced LOCA frequencies of undegraded and/or degraded passive components?

(5) Are the seismic-induced failure frequencies of either piping components or non-piping components expected to be greater than those for non-seismic-induced failures, or are the expected frequencies similar?

(6) What locations within the primary pressure boundary tend to be most susceptible to seismic-induced damage and/or failure?

(7) How can experiential and component testing data be used to bound or develop seismic-induced LOCA frequencies?

(8) Can generic conditional probabilities of degraded component failure be developed as a function of applied seismic loading (or seismic hazard magnitude) for BWR and PWR plants?

(9) What is the range of variability in seismic-induced LOCA frequencies expected among the family of operating BWR and PWR plants?

(10) Does the NRC need to develop criteria for licensees to calculate plant-specific seismic-induced LOCA frequency estimates?

[†] LOCA elicitation findings were for normal operating conditions.

However, for reasons discussed in Section 1.3, the staff decided that explicit determination of seismic-induced failure frequencies for various break sizes would not be feasible. Therefore, after performing a scoping study, the staff shifted the focus of this study to address the following questions:

(1) What is the likelihood of direct failure of unflawed large RCS piping systems resulting from normal operating and seismic loads that would result in breaks larger than the TBS with frequencies greater than 10^{-5} per year?

(2) What would be the critical flaw sizes in the large RCS piping systems associated with stresses resulting from seismic loads with a frequency of occurrence of 10^{-5} per year and 10^{-6} per year?

(3) What is the likelihood of indirect failures (i.e., those attributable to indirect causes, such as support failures) of large RCS piping systems that result in breaks larger than the TBS with frequencies greater than 10^{-5} per year from seismic loads?

While the intent of the above questions (1) and (3) is self evident, it is substantially different than the question of determining seismic-induced frequencies of various break sizes. The focus of these questions is to confirm that the probability of breaks larger than the TBS is less than 10^{-5} per year considering seismic effects. The break size considerations come in via the selection of piping systems to be examined.

In lieu of performing full-scope probabilistic analyses of RCS piping considering various degradation mechanisms to show that the seismic-induced break frequencies of flawed piping are acceptably low, question (2) focuses on the extent of degradation that would have to be present to affect the TBS. In part, this question involves consideration of licensee inspection, repair, and mitigation practices, as well as leak behavior to determine the likelihood of developing critical flaws in service. Thus, question (2) can also be viewed as providing information to preclude the possibility that an existing flaw would grow to critical size before detection and mitigation by certain ISI and leak detection practices. This hybrid deterministic and probabilistic approach can be characterized as a "flaw exclusion" approach to maintain low probabilities of failure.

In order to answer the three aforementioned questions, the staff decided to focus on the following six activities:

(1) Estimate seismic-induced LOCA frequencies for unflawed piping.

(2) Evaluate flawed piping.

(3) Evaluate indirect piping failures.

(4) Review past earthquake experience evaluations.

(5) Review past PRAs.

(6) Review the LLNL study [NRC, 1985a, 1985b, 1985c, and 1988a].

The first three activities provide the primary basis for the assessment and conclusions discussed in detail in Section 4 of this report. The fourth and fifth activities were limited efforts during the scoping phase to gain qualitative and quantitative insights to guide further studies and confirm some of the evaluations performed in first three activities. Sections 2.3 and 2.4 discuss the results of these reviews as they provide background for the entire evaluation. The sixth activity is the review of the LLNL study to estimate the probability of a DEGB in the RCS of the four

U.S. reactor vendor designs. This review has been an integral element of the staff's assessment because its objectives were analogous to those of the current study, and its findings are germane to the issues addressed herein. Also, the LLNL study provides an example of a full-scope probabilistic study, as opposed to the "flaw exclusion" approach that the staff used in the present study. Therefore, Section 3 discusses the LLNL study in greater detail.

2.3 Review of Past Seismic Experience Data

Over the past three decades, numerous strong motion earthquake damage investigation reports have been published. These reports document the findings of Government- and industry-sponsored reconnaissance teams sent to investigate and assess the damage to buildings, industrial facilities, transportation systems, and other major structures immediately following the occurrence of major earthquakes. Organizations involved in these efforts include the NRC, U.S. Department of Energy (DOE), National Institute of Standards and Technology (NIST) of the U.S. Department of Commerce, Electric Power Research Institute (EPRI), Earthquake Engineering Research Institute (EERI), Lawrence Livermore National Laboratory (LLNL), and EQE International. These investigations have provided valuable lessons on seismic response that are applicable to nuclear power plants. However, most of the resultant reports focus primarily on the performance of buildings and major structures, and generally provide only brief, qualitative information on piping system failures.

The present study included a literature review to collect seismic experience data on piping system failures that have occurred as a result of earthquakes at nuclear and fossil power plants and other industrial facilities. As previously discussed, the objective of this review was to collect qualitative and quantitative data that could be used to support and validate the results of the analytical methodology. In particular, this review considered the results of two major NRC-sponsored studies [Stevenson, 1985, 1995], which investigated and provided comprehensive summaries of the performance of piping systems during strong motion earthquakes. This review also took into account additional information from reconnaissance reports [EERI, 2000, 2001, 2003; and Sezen, 2000] for four recent non-U.S. earthquakes. These and the numerous other earthquakes for which damage surveys have been performed over the past 30 years [e.g., Stevenson, 1995] support the same conclusions concerning the performance of welded steel piping systems during actual earthquakes. Specifically, the following factors cause or contribute to failure:

- material degradation (i.e., corrosion and erosion)
- seismic anchor motion (SAM) attributable to large relative motion between buildings or structures from which the piping system is supported
- SAM attributable to large motion of support endpoints (i.e., equipment anchorages) for major components, such as tanks, heat exchangers, pumps, etc.
- interaction of the piping system and its components with structures or components
- interaction from other structures or components, such as structures collapsing on the piping system
- excessive equipment nozzle loads

Seismic experience data also led to the following conclusions:

(1) Overall, the seismic experience data appear to confirm the ruggedness of piping systems and associated components to withstand large earthquake loads beyond the design basis.

(2) The performance of welded steel piping at power plants and industrial facilities has generally been excellent during strong motion earthquakes with peak ground accelerations as high as 0.5g, with a relatively small number of failures observed.

(3) Most reports of piping damage at power plants involved small pipes or tubes, and a significant number of those failures involved only cracks or leaks.

(4) Experience data on pipe supports was collected simultaneously with experience data on piping performance. In general, for above-ground welded-steel piping systems, failure of pipe supports did not lead to failure of piping systems. The failures reported appear to result from support overloads attributable to inadequate seismic restraint or corrosion of supports.

(5) Detailed seismic stress information on piping systems that failed during earthquakes was not obtainable from the available studies and earthquake reconnaissance reports that were reviewed under this effort. However, although such quantitative data were not available, the information collected does provide strong real-world evidence to support the conclusion that properly designed and maintained large-bore piping systems in nuclear power plants have an extremely low probability of catastrophic failure during a strong motion earthquake.

2.4 Review of Probabilistic Risk Assessments (PRAs)

The PRAs conducted for nuclear power plant system performance provide information for evaluation of the following three aspects:

(1) direct failure of RCS piping attributable to the combination of normal operating and earthquake loads

(2) indirect failure of RCS piping attributable to added stresses and deformations induced in the piping system as a result of piping system support failures

(3) indirect failure of RCS piping attributable to added stresses and deformations induced in the piping system as a result of the following causes:

  (a) excessive deformation or failure of in-line components (e.g., valves)

  (b) failure of major components or their supports acting as anchor points for the piping system

In general, in addressing the first aspect, seismic PRAs have not explicitly modeled effects of flawed piping. The previously mentioned NRC-sponsored LLNL study (discussed in detail in Section 3) probabilistically examined the effects of a degradation mechanism on the failure of primary piping. The conclusions of that study (and others) led to the related screening criteria and treatment of the primary system in beyond-design-basis earthquake evaluations. On the second and third aspects of the problem, considerable effort was devoted to its evaluation in the scope of the early seismic PRAs and development of methodologies to address beyond-design-basis earthquakes. The following paragraphs discuss these elements in greater detail.

Lessons learned from completed and submitted seismic PRAs fall into two time frames. The first time frame precedes the submittal of the numerous responses to the NRC's IPEEE program. This time frame included fragility function development for structures, systems, and components (SSCs) in a wide range of nuclear power plants of all types (the data represented more than 20 seismic PRAs). These data, together with insights from generic fragility function development as part of the NRC's Seismic Safety Margins Research Program (SSMRP), and conclusions drawn from the detailed LLNL studies of DEGB and indirect DEGB (summarized in Section 3), led to the development of the Seismic Margin Assessment (SMA) methodology [EPRI, 1991], and the following general conclusions:

- Piping and other passive components of the RCS generally exhibit very high seismic capacities.

- Valves, piping, and pipe supports have a significantly higher capacity than other RCS components, although there is a wide range in the reported values. As known from the SMA methodology, valves may have seismic vulnerability depending on their configuration, type, and function to be performed.

- Components in the RCS are more vulnerable than the piping and pipe supports. Total or partial component failure, such as component support failure, could lead to additional stresses on RCS piping and enhance the potential for failure.

These observations and conclusions also led to the screening criteria of the SMA methodology. For the portions of the RCS of interest to this study, no explicit evaluation of the nuclear steam supply system (NSSS) is required for review level earthquakes of approximately 0.5 peak ground acceleration, with the possible exception that pressurizer supports should be evaluated for PWRs, and reactor vessel supports and recirculation pump supports should be evaluated for BWRs. This guidance was implemented in the NRC's IPEEE program for SMA and seismic PRA methodologies.

The second time frame corresponds to evaluation of licensees' submittals in response to the NRC's IPEEE program. Two studies summarized the results of these licensee submittals in response to Supplement 4 to the NRC's Generic Letter 88-20 [NRC, 1988b], which asked each licensee to perform an IPEEE evaluation to identify vulnerabilities, if any, to severe accidents and report results together with any licensee-determined plant improvements and corrective actions. An EPRI study [EPRI, 2000] examined the results for seismic events only; whereas, the NRC study, NUREG-1742 [NRC, 2001], considered all manner of external events. Of interest to this study are the conclusions drawn from the seismic PRAs.

> **EPRI Review [EPRI, 2000].** EPRI's review considered 28 of the 75 seismic PRA submittals, representing 41 units. The remaining submittals and units used either the EPRI SMA methodology, the NRC's seismic margin methodology, or a more site-specific approach. Selected multi-unit sites with shared systems and functions performed seismic PRAs for the combined plant. (This is why the 28 submittals reviewed by EPRI represented a total of 41 units.) Of relevance to the present study, the EPRI review includes summary tables with information on IPEEE seismic PRA dominant risk contributors (systems and components). In four cases, the EPRI review identified the NSSS as a dominant risk contributor. However, upon more detailed evaluation, the NSSS was considered a dominant risk contributor in only one of the four cases, in which the NSSS failure was represented by a surrogate element (i.e., no specific models were developed for this system because it was screened out with high capacity). In the other three cases, it appears that the dominant contributors were actually

structural failures, which led to some system failure or heat exchanger failure for decay heat removal. Only the case of heat exchanger failure possibly falls within the scope of interest in the present study.

NRC Review [NRC, 2001]. The NRC's review of the seismic PRA submittals considered 27 of the 70 submittals reviewed, representing 40 units. As with the EPRI review (above), the remaining submittals and units used either the EPRI SMA methodology, the NRC's seismic margin methodology, or a more site-specific approach. Selected multi-unit sites with shared systems and functions performed seismic PRAs for the combined plant. (This is why the 27 submittals reviewed by the NRC represented a total of 40 units.) For the most part, the observations and conclusions drawn from this review mirror those of the EPRI study [EPRI, 2000], with the exception that the NRC's review did not explicitly identify the NSSS as being a dominant risk contributor in any of the seismic PRA results. However, in one case, the NRC's review identified the high-pressure injection system in a BWR as being a dominant contributor (but without detail as to which components or failure modes dominated). It is highly unlikely that this case involved piping failure, although it might represent a component failure (likely non-structural in origin). The same conclusion can be drawn from the NRC's review as the EPRI review. Specifically, for the seismic PRAs reviewed, the only potential contributor to risk is major component failure, which could lead to additional stresses in piping systems and subsequent failure at high earthquake excitation levels.

The conclusion from the 25 years of evaluations of beyond-design-basis earthquakes for nuclear power plants is that the only NSSS failure modes identified to be of potential interest are attributable to failures of major components or their supports. In the present study, indirectly induced failure of NSSS piping attributable to major component support failures was considered on a pilot plant basis, as reported in Section 4.6.

3 SUMMARY OF THE LLNL STUDY

3.1 Objective

As discussed in Section 2.2, the NRC contracted with LLNL in 1980 to estimate the probability of a DEGB in RCS piping. The purpose of this study, conducted as a part of the load combination program at LLNL, was to provide the NRC with the technical basis for the following activities:

- Reevaluate the then-current general design requirement that DEGB could be assumed in the design of nuclear power plant structures, systems, and components as the basis for considering the effects of a postulated pipe break.

- Determine whether an earthquake could induce a DEGB, and thus reevaluate the design requirement that pipe break loads be combined with loads resulting from an SSE.

- Make licensing decisions concerning the replacement, upgrade, or redesign of piping systems, or addressing such issues as the need for pipe whip restraints on RCS piping.

The following excerpt from the NRC report [NRC, 1985a] further highlights why this was important:

> *Elimination of DEGB as a design-basis event for PWR RCS piping could have far-reaching consequences. If it can be shown that an earthquake will not induce DEGB, then the two can be considered independent random events whose probability of simultaneous occurrence is negligibly low; thus, the design requirement that DEGB and SSE loads be combined could be removed. If the probability of DEGB is very low under all plant conditions, including seismic events, then asymmetric blowdown loads in PWR plants could be eliminated. Reaction loads on pipe and component supports could be reduced. Jet impingement loads, as well as environmental effects due to a LOCA could be modified accordingly. Pipe whip restraints could be eliminated altogether, as without a double-ended break, the pipe would retain at least geometric integrity. This last benefit would apply to operating plants as well as those in design or under construction, because once removed for periodic weld inspection, pipe whip restraints would not have to be reinstalled.*

In order to achieve the objective of estimating the probability that a DEGB occurs in the RCS piping of light-water reactors, the LLNL study considered two potential causes for DEGB:

- fatigue crack growth at welded joints resulting from the combined effects of thermal, pressure, seismic, and other cyclic loads

- seismic-induced failure of component supports or other equipment whose failure would in turn cause an RCS pipe to break (indirect DEGB)

3.2 Overall Scope and Approach

To arrive at a general conclusion about the probability of DEGB in the RCS piping of PWR plants, LLNL took a vendor-by-vendor approach. For each of the three PWR vendors (Westinghouse, CE, and B&W), the specific objectives were as follows:

(1) Estimate the probability of a DEGB, taking into account such contributing factors as initial crack size, pipe stresses attributable to normal operation and sudden extreme loads (e.g., earthquakes), the crack growth characteristics of pipe materials, and the capability to nondestructively detect cracks or to detect a leak if a crack penetrated the pipe wall.

(2) Estimate the probability of an indirect DEGB by identifying critical component supports or equipment of which failure could result in a pipe break, determining the corresponding seismic "fragility" (i.e., relationship between seismic response and probability of failure), and combining this result with the probability of an earthquake that produces a certain level of excitation ("seismic hazard").

(3) For both causes of a DEGB, perform sensitivity studies to identify key parameters contributing to the probability of pipe break.

(4) For both causes of a DEGB, perform uncertainty studies to determine how uncertainties in input data affect the uncertainty in the final estimated probability of pipe break.

Although, the overall objectives and approach of the BWR study were generally the same, the LLNL study, documented in NUREG/CR-4792 [NRC, 1988a], noted the following two considerations for BWRs:

- susceptibility of certain BWR stainless steels to stress corrosion cracking, which required the development of an advanced probabilistic model of corrosion phenomena (note that, at the time, stress corrosion was not perceived as a problem in PWR primary loop piping and, therefore, was not considered in the evaluation)

- greater complexity and flexibility of BWR reactor coolant piping, compared to PWR primary loops, which required incorporating conventional supports (e.g., snubbers, spring hangers) for piping and light loop components in the evaluation

The scope of the PWR piping was limited to large RCS piping. For example, for CE plants, the RCS piping considered included one hot leg, two cold legs, and two suction crossover legs connecting the reactor pressure vessel, one steam generator, and two reactor coolant pumps; however, the study did not include the pressurizer surge line. For BWRs, the study considered recirculation loop piping, main steam line piping, and main feedwater piping.

For Westinghouse and CE evaluations, LLNL designated a single reference (or "pilot") plant, as a basis for developing the methodology and for extensive sensitivity studies to identify the influence that individual parameters have on DEGB probabilities. Thus, each pilot plant was used to develop and validate the assessment methodology that was later applied in the corresponding generic study for each vendor.

In the generic study of RCS piping manufactured by each NSSS vendor, LLNL evaluated individual plants or group of plants sharing certain common or similar characteristics. Thus, the results (that is, estimates of the annual probability of DEGB, including uncertainty bounds) applied to the single pilot plant or members of the group, as appropriate, and were interpreted as characteristic of that pilot plant or group. Thus, the generic evaluation represented a "production" application of the assessment methodology. LLNL then performed a lesser scope analysis for B&W plants. Additional details of the BWR scope appear in NUREG/CR-4792 [NRC, 1988a].

In estimating the probability of a direct DEGB for PWRs, LLNL only considered fatigue crack growth from the combined effects of thermal, pressure, seismic, and other cyclic loads as the mechanism leading to pipe break or leak. That is, LLNL did not consider hydrodynamic loads attributable to water hammer, because such loads had never been observed in PWR RCS piping. Similarly, LLNL excluded intergranular stress corrosion cracking (IGSCC) because such problems had not been noted in ferritic pipe materials.

To estimate the probability of an indirect DEGB, LLNL considered the safety margins against seismic failure for critical components of which failure could in turn cause a reactor coolant pipe to break. By combining this information with seismic hazard curves (in terms of peak ground acceleration), LLNL was able to estimate the annual probability of a guillotine break caused by earthquakes. In particular, LLNL used generic seismic hazard curves (including uncertainty) for plants east of the Rocky Mountains, as shown in Figure 3-1. By contrast, LLNL evaluated west coast plants using the then-available site-specific information.

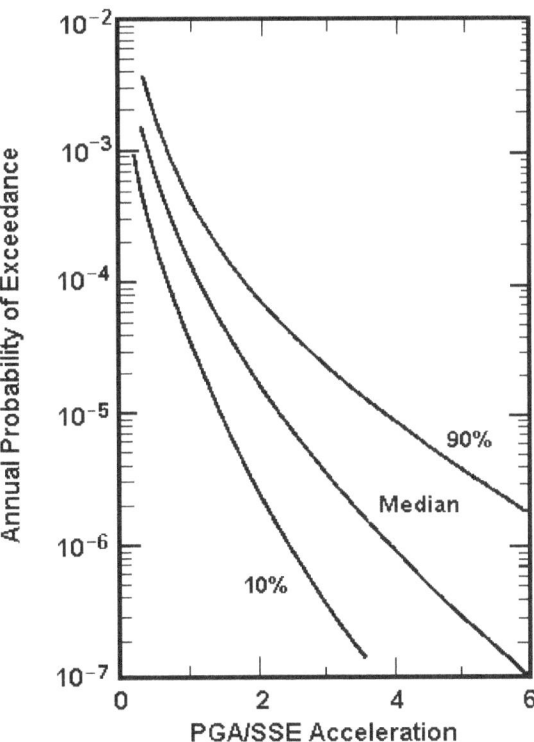

Figure 3-1 Generic Seismic Hazard Curves for Plants East of the Rocky Mountains

3.2.1 Approach: Direct DEGB

To account for the randomness of events and parameters associated with plant operation, LLNL used a probabilistic fracture mechanics methodology that involved estimating crack growth and assessing crack stability during the lifetime of the plant. This section is a condensed version of the methodology overview given in NUREG/CR-3663, Vol. 2 [NRC, 1985b], as illustrated in the simplified flow chart in Figure 3-2. (In that figure, the left column shows the analytical process, while the right column shows the necessary input information and various simulation models used for each step of the analytical process.)

In the LLNL approach, the analytical process is divided into two parts. The first involves calculating the conditional probability of a leak or DEGB at individual weld joints, assuming that a crack exists at the given joint, the plant experiences various loading conditions at any given time, and a seismic event of a specific intensity occurs at a specific time.

The second part of the process, known as "system failure" probability analysis, involves estimating the probability of a leak or DEGB for the entire RCS piping system, considering all of the associated weld joints. "System failure" is defined as a leak or DEGB (also called a "pipe failure") occurring in at least one of the weld joints of a reactor coolant loop (RCL) during the lifetime of the plant. As such, system failure probabilities are presented on an annual basis. However, it is important to note that the system failure analysis was actually carried out for the entire duration of plant life (assumed to be 40 years), and the system failure probabilities are not necessarily uniform over this long duration. Thus, the LLNL approach began with estimating the failure probability of a weld joint and then determined the system failure probability of the RCS piping.

For each weld joint of the piping system, LLNL used a Monte Carlo simulation technique to calculate the conditional leak or DEGB probability at any specific time in the life of the plant. In so doing, LLNL subjected each weld joint to a stress history associated with plant events, such as normal heatup or cooldown, anticipated transients, and occurrence of potential earthquakes.

In this process, LLNL began the simulation with the random selection of sample crack sizes from a sampling space, followed by calculation of conditional probabilities associated with the selected crack sizes. LLNL then applied fracture mechanics theory to calculate the growth of the cracks and determine whether the pipe fracture (i.e., either leak or DEGB) will occur as the cracks grow during the lifetime of the plant. In so doing, LLNL simulated various parameters related to crack and leak detection, such as, preservice inspection, hydrostatic proof test, and leak detection (Figure 3-2).

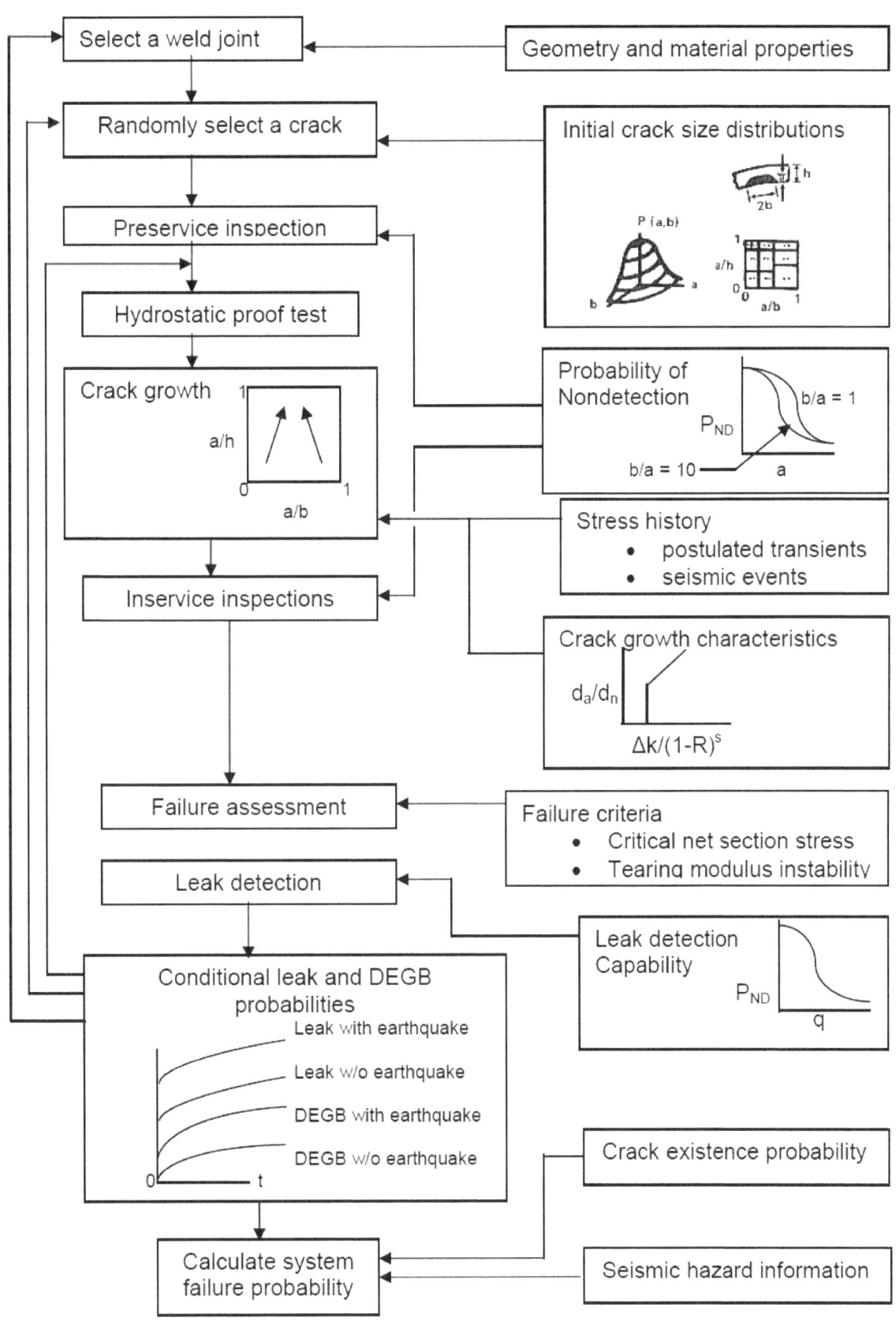

Figure 3-2 Flow Chart for Probabilistic Assessment of Piping Integrity

Fatigue crack growth takes into account the cyclic stress history of various thermal transients and postulated seismic events. The failure criteria applied involve either the critical net section collapse approach or the tearing modulus instability approach, depending upon their applicability to the material characteristics and geometric conditions of the pipe. The stress state of the plant varies as the various loading events occur throughout the plant life. Therefore, it was necessary to monitor or calculate the state of the cracks, considering the effects of these loading events as time progresses. The time of occurrence of these loading events can be either deterministic or stochastic. The seismic events were treated as stochastic and assumed to be describable by a Poisson process in calculating the system failure probability. As previously stated, the LLNL study used a generic hazard curve (Figure 3-1) for sites east of the Rocky Mountains. Other plant transients were considered uniformly spaced throughout the life of the plant.

Inservice inspections were neglected in the LLNL study, because ISI programs vary greatly from plant to plant. The frequency of transient events were based on the postulation used in the plant design and were considered to be conservative.

The LLNL study assessed the effect of an earthquake of specific intensity on the failure probability at each weld joint at specific times during the plant life. First, the probability of failure was determined with no seismic events. Then, earthquake loads of specified intensity, usually expressed in terms of peak ground acceleration (PGA), were imposed on normal operating conditions, and the increase in failure probability attributable to the earthquake was added to the failure probability. This process was repeated for a wide range of earthquake intensities.

The above calculation procedure yields the conditional leak or DEGB probabilities (conditioned on both the existence of a crack and the occurrence of an earthquake of given size) as a function of time for a specific weld joint. LLNL repeated this analytical process for all welds in a single RCS loop, and then estimated system failure probability based on the assumption that the multiple loops within a plant are identical in geometry and have an identical stress history at each corresponding weld joint.

In estimating system failure probabilities, LLNL examined the following three scenarios:

(1) one or more earthquake(s) during plant life and a failure occurring simultaneously with the first earthquake

(2) one or more earthquake(s) during plant life and a failure occurring before the first earthquake

(3) a failure and no earthquake during plant life

The LLNL study also included detailed uncertainty analysis, considering aleatory and epistemic uncertainties. Details can be found in NUREG/CR-3660 [NRC, 1985a].

3.2.2 Approach: Indirect DEGB

In evaluating indirect DEGB probability, LLNL assumed that the failure of any one of the major RCL component supports would result in a DEGB. Also, for a limited sample, LLNL considered the failure of an overhead crane because such a failure and its possibility of impacting the RCS were judged to be credible. For sample plants, LLNL first identified critical components for which seismic fragility was determined. In order to select the governing plants for detailed consideration, LLNL generated a consistent determination of the seismic capacities and SSE factors of safety for all plants in the evaluation, and then calculated the probability of indirectly induced failure leading to DEGB for selected plants. Finally, LLNL estimated the non-conditional probability of an indirect DEGB by statistically combining seismic hazard curves with a "plant-level" fragility derived from the individual component fragilities.

The LLNL study noted that the results suggested that seismic-induced support failure is the weak link in the DEGB evaluation, and the strength of component supports, currently designed for the combination of SSE plus DEGB, should not be reduced.

3.3 Key Findings

3.3.1 Findings: Direct DEGB

Overall, the LLNL study concluded that the probability of a direct DEGB in RCS piping is very low for both PWR and BWR plants. LLNL further concluded that the probability of a leak or direct DEGB in the RCS piping is negligibly affected by earthquakes, to the extent that direct DEGB and earthquake can be considered independent random events. The following is a summary of key results reported in the LLNL study:

- For Westinghouse plants (east and west of the Rocky Mountains), the median probability of a direct DEGB in RCS piping is on the order of 10^{-10} event per year, while the median probability of a leak (through-wall crack) in RCS piping is on the order of 10^{-7} event per year.

- For CE plants, the best-estimate probability of a direct DEGB in RCS piping is on the order of 10^{-13} event per year, while the median probability of a leak (through-wall crack) in RCS piping is on the order of 10^{-8} event per year.

- For B&W plants, a limited evaluation of RCS piping in one B&W plant supported the conclusion that the probability of a direct DEGB in RCS piping should be similar to that of Westinghouse plants (on the order of 10^{-10} event per year).

3.3.2 Findings: Indirect DEGB

The LLNL study concluded that indirect causes are clearly the more dominant mechanism leading to a DEGB in RCS piping. The probability of an indirect DEGB is a strong function of seismic hazard, but is nonetheless low even when earthquakes significantly greater than the SSE are considered. Table 3-1 shows sample results.

Table 3-1 Annual Probabilities of Indirect DEGB for Events Per Year

Group A Plants **(Combustion Engineering)**	Confidence Limit [1]		
	10%	50%	90%
Calvert Cliffs	2.3×10^{-8}	6.1×10^{-7}	6.1×10^{-6}
Millstone 2	9.0×10^{-10}	6.6×10^{-8}	1.2×10^{-6}
Palisades	5.0×10^{-7}	6.4×10^{-6}	5.2×10^{-5}
St. Lucie 1	1.2×10^{-8}	3.8×10^{-7}	4.1×10^{-6}
St. Lucie 2	6.6×10^{-8}	1.4×10^{-6}	1.1×10^{-5}
Westinghouse Lowest Capacity Plant	2.3×10^{-7}	3.3×10^{-6}	2.3×10^{-5}

(1)   A confidence limit of 90% implies that there is a 90% subjective probability (confidence) that the probability of indirect DEGB is less than the indicated value.
(2)   Generic seismic hazard curves (Figure 3-1) used in evaluation.

With respect to indirect failure, the LLNL study yielded the following results:

- For Westinghouse plants, the median probability of an indirect DEGB in RCS piping is about 10^{-7} event per year for plants east of the Rocky Mountains and 10^{-6} event per plant for west coast plants.

- For CE plants, the median probability of an indirect DEGB in RCS piping is about 10^{-6} event per year for older plants and less than 10^{-8} event per year for newer plants.

- For B&W plants, a limited evaluation of RCS piping was performed for two B&W plants. The probability of an indirect DEGB in one plant was 10^{-10} event per year, and in the other plant it was 10^{-7} event per year.

- For BWR plants, the median probability of an indirect DEGB is estimated to be 10^{-8} event per year.

With respect to indirect failure, it is important to note the following recommendation from the LLNL study:

> *...seismically induced support failure is the weak link in the DEGB evaluation, we further recommend that the strength of component supports, currently designed for the combination of SSE plus DEGB, not be reduced.*

3.4 Relationship of the LLNL Study to the Present Effort

Results of the LLNL study are relevant to the present effort and provide a degree of assurance that probabilities of seismic-induced large breaks can be shown to be low. The LLNL study has directly formed the basis for the staff's present approach related to indirect failure, as discussed in Section 4.6. However, because of the following considerations, this study (by itself) is not used as a sole technical basis for the present effort:

(1) There have been considerable advancements in the field of seismology, particularly in the area of source modeling and ground motion estimates. For the central United States and east coast nuclear power plant sites, new ground motion models coupled with recent earthquake events have exhibited ground motion characteristics that are significantly different than those used in the LLNL study. While, it is very likely that the use of new hazard estimates will also result in low failure probabilities, this is best demonstrated by performing additional analyses.

(2) For PWRs, the LLNL study calculated crack growth of as-fabricated surface flaws attributable to fatigue resulting from the combined effects of thermal, pressure, seismic, and other cyclic loads. Since then, stress corrosion mechanisms, such as primary water stress-corrosion cracking (PWSCC), have emerged.

(3) The fully probabilistic fracture mechanics analysis of piping with stress-corrosion cracks is a very difficult undertaking (for both BWRs and PWRs). LLNL study used a computer code developed for such analyses in the 1980s as part of its study. Since then, significant advances have occurred in probabilistic fracture mechanics to predict the lengths and multiple initiations of stress corrosion cracks. The length of the surface crack is the most important consideration in determining break probabilities. To address this issue, RES has an ongoing separate developmental effort to make more realistic improvements in LOCA-related probabilistic fracture mechanics predictions with validations from past history of cracking.

The LLNL approach is the most comprehensive type to explicitly calculate seismic-induced break frequencies, and can be used in conjunction with the new hazard curves and with incorporation of improvements in probabilistic fracture mechanics to more precisely account for applicable degradation mechanisms.

4 THE STAFF'S APPROACH FOR THE PRESENT STUDY

4.1 General Approach and Scope of Analyses

As discussed in Section 1.3, the staff's approach and the scope of the present study were determined, in part, by the following factors:

- highly site-specific and plant-specific frequencies of seismic-induced LOCAs
- ongoing and recent developments in seismic hazard studies (for example, as part of early site permit activities)
- availability of plant-specific information such as normal operating stresses, design seismic stresses, and material properties

These factors dictated the number and type of plants that could be analyzed and the hazard curves chosen for the present study.

In addition, as discussed in Section 2.2, for various reasons and stated objectives, a full-scope probabilistic approach similar to that used in the LLNL study was neither necessary nor feasible.

In the staff's approach, the starting point was the availability of design information. The best source available to the staff for plant-specific design basis was the leak-before-break (LBB) analyses previously submitted by licensees. This database (described in Section 4.2) gives information on normal operating and SSE seismic stresses for pipe systems of interest, as well as information such as pipe dimensions and material properties. Because the LBB database contains information for PWR plants only, and similar information was not available for BWR plants, the staff only analyzed PWR plants in this study. However, the approach used in this study should be equally applicable to BWR plants.

Seismic stresses and seismic-induced break frequencies are direct functions of estimated seismic hazard at the site. Uncertainties in seismic hazard generally dominate uncertainties in the seismic risk assessments. For this study, the staff selected the update of the so-called revised LLNL hazard curves [Sobel, 1994]. Reasons for and implications of this choice are discussed in Section 4.3. The LLNL hazard curves are available for 69 sites east of the Rocky Mountains; therefore, the staff focused on PWRs located in the central and eastern United States.

The staff primarily concentrated on large-diameter RCS piping, such as the hot leg, cold leg, and cross-over legs, which have larger diameters than the proposed TBS.

For the present study, the staff used a hybrid deterministic and probabilistic approach, which included some sensitivity studies to address uncertainties. The evaluations included the following key elements:

- Stresses attributable to dead load, pressure, and thermal loading conditions were taken as point estimates from the LBB database. Where detailed information was not available (e.g., the breakdown of stresses attributable to thermal loading conditions and dead load and pressure), the staff made assumptions and, if deemed appropriate, performed sensitivity studies on the parameters.

- The evaluation of seismic stresses for higher earthquake levels was based on the SSE stresses provided in the LBB database. However, the staff adjusted the SSE stresses for conservatism, or nonconservatism, to account for ground motion and soil/structure analysis parameters and procedures, as well as piping system analysis parameters and procedures. The staff then extrapolated these "best estimate" SSE stresses to higher earthquake levels as point estimates for the direct seismic-induced break evaluation. The higher earthquake levels of interest to this study were PGAs with annual exceedance probabilities of 10^{-5} and 10^{-6}. These earthquake levels were defined by the LLNL mean seismic curves at each plant site of interest. Adjustment for conservatism, or nonconservatism, in the seismic stresses and extrapolation to high earthquake levels was achieved by implementing the seismic PRA Scale Factor approach (described in detail in Section 4.4.2), in which the Scale Factors are inverse of Factors of Safety described in Section 4.4.2.2, and are median values that correct the seismic stresses for the above-mentioned effects.

- Indirectly induced breaks attributable to seismic events were evaluated on a limited basis, following the LLNL approach described in Section (3), and considering two PWR plants (one Westinghouse and one CE). Very limited information was available regarding the design details of these two plants. Hence, the staff used the information provided in [NRC, 1985a, 1985b, 1985c, and 1988a], supplemented by the LLNL seismic hazard data [Sobel, 1994]. In the context of the Scale Factors approach outlined above, the staff considered median values and full uncertainties. The staff then performed the resulting analyses by integrating over each plant site's seismic hazard from low levels to an upper limit of 1.5g PGA. Section 4.6 presents the approach and results of the analyses.

- Strength and load resistance parameters were based on mean material properties in the flawed pipe evaluations. For unflawed piping, the allowable stresses from Section II of the ASME BPV Code were used to ensure consistency with the failure criterion used.

- Sensitivity analyses were performed for a number of issues, as discussed in the subsequent sections.

The staff's approach yielded results, which are characterized as "best estimates" of the critical flaw size in the flawed piping analysis.

From the LBB database, the staff selected 27 PWRs to cover representative operating, seismic, and total stresses; a variety of pipe and weld materials with varying toughness properties; and a range of seismic hazard. In so doing, the staff focused the study on PWRs that are located on rock sites, because examination of the LBB database revealed that, as anticipated, higher seismic stresses in piping systems generally occurred at rock sites especially when extrapolated to the 10^{-5} per year seismic hazard. Coincidently, it was plausible to evaluate the large number of rock founded units implementing the Scale Factor method (Section 4.4.2), because much of the important seismic response information was available with minimal effort. The staff also considered three plants founded on soil of varying characteristics.

In summary, the above considerations resulted in the following scope:

- 27 PWRs (24 on rock sites and 3 on soil sites)
- PWRs only (Westinghouse and CE)
- use of seismic hazard reported in NUREG-1488 [Sobel, 1994], which contains 1993 LLNL seismic hazard curves (Appendix A) and Uniform Hazard Spectra (UHS) (Appendix B)
- plants located east of the Rocky Mountains
- large-diameter RCS piping systems, of which failures may affect the TBS

Despite the scope of this particular study, it should be noted that the staff's approach is not limited to PWRs or plants located in the central and eastern United States. The same approach can be applied to BWR piping systems and plants located in the western United States.

After describing the LBB database and considerations for using the LLNL hazard curves, Sections 4.3 through 4.6 describe approaches and results of analytical activities related to three main activities outlined in Section 2.2: (1) unflawed piping, (2) flawed piping, and (3) indirect failures.

4.2 Description of the LBB Database

In conducting these evaluations, it was necessary to have available the normal operating stresses and SSE stresses for the pipe systems of interest, as well as some key information such as the pipe dimensions and materials. This information was available as part of the database initially created in NUREG/CR-6765 [Scott, 2002] and referred to here as the LBB database.

The LBB database was in spreadsheet format and contained information from licensees' LBB submittals to the NRC. These submittals were initially compiled by the NRC in December 1987 and were in hardcopy form; however, the information from the hardcopy LBB submittals was subsequently updated and keyed into the LBB database spreadsheet.

The implementation of the LBB database involved reviewing the number of plants and documents that were available for review. Because BWRs were not approved for LBB relief from dynamic loads for the full-break assumption in the initial plant design basis, the information available to consider in this review was mainly from PWRs.

The hardcopy LBB submittals contained the following proprietary information:

- submittals for several units on the same site
- plant-specific submittals that also contained submittals for several other "identical" plants
- submittals for advanced reactors that have not been built in the United States
- submittals for several different piping systems from the same plant

For purposes of this evaluation, it was necessary to process the hardcopy submittals as follows. Individual document and data were sorted by unit and piping system. Extensive information on all primary piping systems was available. Virtually all primary piping systems in PWR plants (hot legs, cold legs, and cross-over legs) have been approved for LBB relief from dynamic loads for the full-break assumption in the initial plant design basis. This relief could entail eliminating pipe whip restraints or jet impingement shields. In addition, smaller diameter piping systems, such as surge lines, residual heat removal (RHR) lines, and other safety injection system lines, have been approved for LBB in some plants and, hence, information was also available for these systems.

The current LBB database, as updated in this study and compiled in multiple spreadsheets, consists of submittals from 56 units with information on 177 pipe systems. The database consists of two parts. The first part contains basic information and nonconfidential or proprietary data for each plant, as follows:

(1) Plant name
(2) Docket number
(3) Utility
(4) Type
(5) Vendor
(6) AE firm
(7) Power (megawatt)
(8) Plant operating date
(9) LBB status
(10) LBB summary information:
   (a) Pipe system
   (b) Date of submittal
   (c) Date of approval
   (d) Number of documents reviewed
   (e) Restrictions on the LBB approval
   (f) Key references
   (g) NRC requests for additional information

In addition, comments may be embedded to provide explanatory notes for the data.

The second part of the database contains proprietary information from the LBB applications and, hence, cannot be publicly released. This portion of the database contains the following columns:

(1) Plant Name
(2) Type (i.e., PWR)
(3) Vendor
(4) LBB analysis organization
(5) LBB Status
(6) LBB Application Technical Information
(7) Pipe system
(8) Inside diameter (inches)
(9) Thickness (inches)
(10) Material
(11) Critical location
(12) Normal operating pressure (psig)
(13) Normal operating temperature (°F)
(14) Normal operating stress assumptions (dead-weight, thermal expansion, and pressure)
 (a) F_x (lbs)
 (b) F_y (lbs)
 (c) F_z (lbs)
 (d) M_x (ft-kips)
 (e) M_y (ft-kips)
 (f) M_z (ft-kips)
(15) SSE assumptions
 (a) F_x (lbs)
 (b) M_x (ft-kips)
 (c) M_y (ft-kips)
 (d) M_z (ft-kips)
(16) Leak-rate analysis assumptions and code used
(17) Fracture analysis assumptions and code used
(18) Source of elastic-plastic fracture toughness (J-R curve),
(19) Stress-strain curve used
(20) Margins calculated by submitter
(21) Margins calculated by NRC staff
(22) Comments (three different columns for various types of comments)

As with the "Basic Information" portion of the data, explanatory notes with expanded technical information are included.

As previously stated, the staff selected 27 PWRs (mostly on rock sites) for this study. The basis for the selection was to cover representative cases of different materials, plants with high and low normal stresses, and plants with high and low seismic stresses. A wide variation in seismic hazard was also reflected in this selection.

4.3 Seismic Hazard

As previously stated, the staff chose to use the updated LLNL seismic hazard curves and UHS [Sobel, 1994] for evaluations in this study. These seismic data are the more recent publicly available data for the 69 nuclear power plant sites in the central and eastern United States.

The staff recognizes that there has been a significant evolution in the data collected and processed, the seismic hazard assessment methodology, and their implementation since the publication of the LLNL seismic hazard [Sobel, 1994]. This is most clear from the studies conducted for three sites for early site permit applications. Their submittals have highlighted some of the differences between the current results and those of a decade ago. These early site permit hazard studies are currently under review, and the industry has undertaken additional studies. Results of a U.S. Geological Survey study are also under review. The staff also recognizes the state of flux of the seismic hazard assessment process, as well as the lack of publicly available, alternative, extensive data for the 69 nuclear power plant sites, and considers the LLNL seismic hazard as the best available choice for this effort. Furthermore, insights obtained from the use of the LLNL seismic hazard will still be valid. The overall procedure is viable for any set of seismic hazard data (i.e., for an individual site or a larger set). However, the staff recognizes recent efforts underway to reevaluate the seismic hazard and the possibility of potential impact on some sites.

Figure 4-1 shows the mean LLNL seismic hazard curves for 43 PWR plant sites, which the staff used in this study to determine seismic stresses at various levels to compare with the failure criterion and determine critical flaw sizes. The staff also used the Weibull distributions to fit each of these curves for calculation of the probability of exceedance of PGA values.

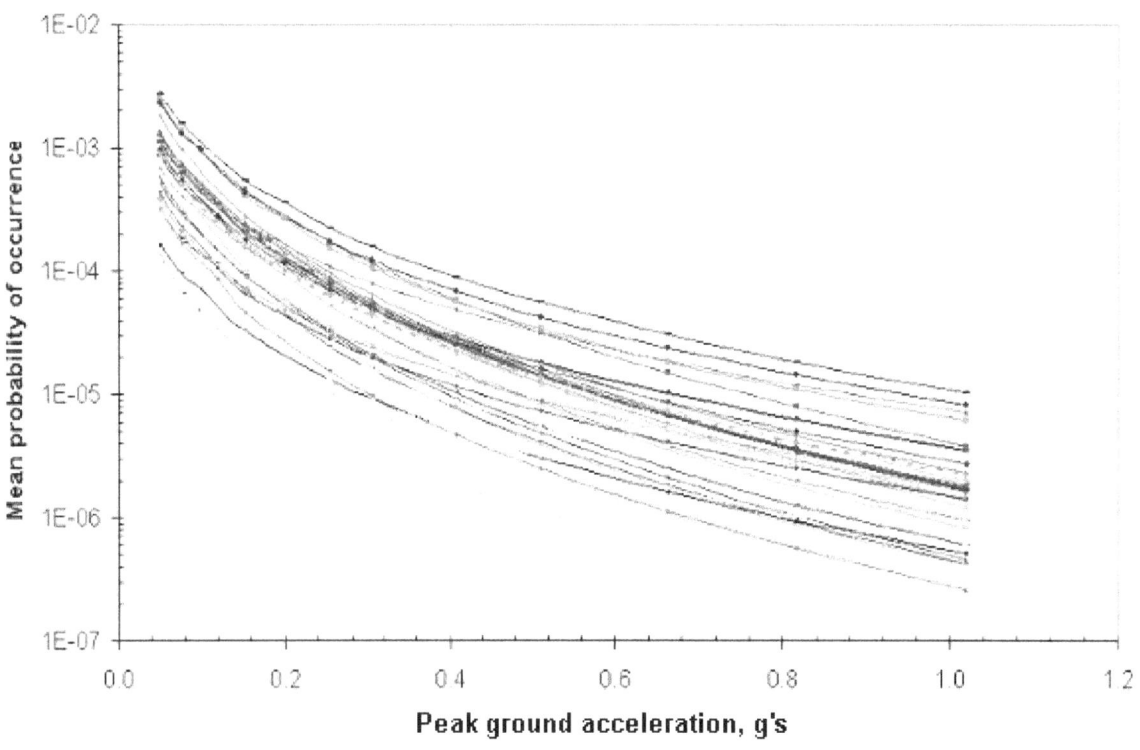

Figure 4-1   Seismic Hazard Curves for 43 PWR Plant Sites East of the Rocky Mountains

4.4 Unflawed Piping

4.4.1 Analysis – Approach and Key Steps

The staff's objective in evaluating the unflawed piping (as defined in the footnote* in Section 2.1) was to determine whether considering seismic loading conditions significantly increases the probability of a break above the TBS.

For a given piping system in a given plant, the process is outlined in Figure 4-2. Steps involved in boxes of Figure 4-2 are briefly explained below with an example. The basis and key assumption involved are discussed in detail in subsequent sections. Results and assessment of all cases analyzed are discussed in Section 4.4.4.

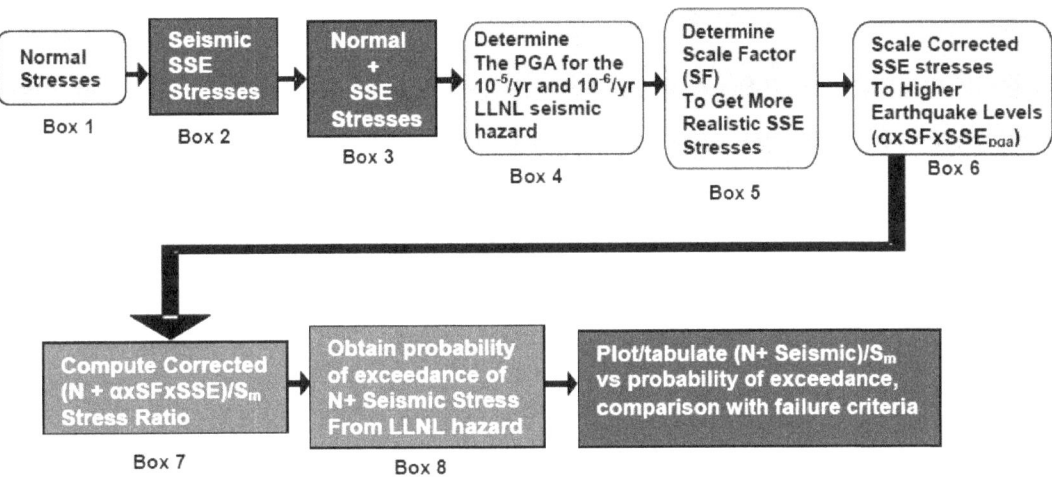

Figure 4-2 Approach and Key Steps for Unflawed Pipe Evaluation

The process begins with selection of a plant from the LBB database from which piping systems of interest are selected. The LBB database gives the maximum normal stress, SSE stress, and combined normal and SSE stress. Material properties are also available from the database. The SSE PGA values come from plant safety analysis reports. Table 4-1 provides such information for the cold leg of a hypothetical PWR, called Plant A. Seismic parameters and pipe stresses are based on an average of a sample of 27 PWR plants in the LBB database.

Table 4-1   Information from the LBB Database

Plant	S_m (ksi)	Maximum Normal Stress, N (ksi) Box 1	Maximum SSE (ksi) Box 2	N+SSE Stress (ksi) Box 3
Plant A	18.8	9.20	9.89	19.09

As seen in Table 4-1:
Normal stress, N = 9.20 ksi (Box 1)
SSE stress = 9.89 ksi (Box 2)
Normal + SSE stress = 19.09 ksi (Box 3)
These stresses are located at a nozzle weld.

The design-basis analyses are conservative processes. For the purposes of this study, more realistic estimates of seismic stresses were needed. Therefore, Scale Factors were developed to estimate extrapolated stresses at earthquake levels corresponding to annual probabilities of exceedance of 10^{-5} or 10^{-6} depending on the analysis to be performed. These Scale Factors adjust SSE stresses from design to best estimate values and extrapolate these best estimate values to higher earthquake levels, taking into account realistic ground motion descriptions and changes in the soil-structure-piping system properties as excitation levels increase. Using the procedure outlined in Section 4.4.2.2:

Scale Factor (SF) for Plant A = 0.64

Table 4-2 provides the information necessary to estimate combined stresses at various earthquake levels and their frequency of occurrences corresponding to steps outlined in Boxes 6, 7, and 8 of Figure 4-2.

Table 4-2 Estimates of Normalized Stress Ratios and Probability of Exceedance

Hazard Curve for Plant A			$\alpha =$ ACC./SSE$_{PGA}$	CORRECTED SEISMIC STRESSES, ksi $\alpha \times SF \times SSE$	(N + Seismic)/S$_m$ VALUES
ACCELERATION		PROBABILITY OF EXCEEDANCE			
(cm/sec^2)	g				
50	0.051	1.21E-03	0.31	1.99	0.59
75	0.076	6.79E-04	0.47	2.98	0.65
150	0.153	2.17E-04	0.94	5.96	0.81
250	0.255	7.99E-05	1.57	9.93	1.02
300	0.306	5.39E-05	1.88	11.91	1.12
400	0.408	2.76E-05	2.51	15.88	1.33
500	0.509	1.57E-05	3.13	19.86	1.54
650	0.662	7.67E-06	4.07	25.81	1.86
800	0.815	4.15E-06	5.01	31.77	2.18
1000	1.019	2.04E-06	6.26	39.71	2.60
1200	1.223	1.10E-06	7.52	47.65	3.02
1400	1.426	6.29E-07	8.77	55.60	3.44
1600	1.630	3.78E-07	10.02	63.54	3.86

The first three columns tabulate data from the LLNL 1993 mean seismic hazard curve for Plant A in terms of PGA, showing various PGA levels and corresponding probabilities of exceedance. The data in Column 3 reflects an average of mean seismic hazard curves of 27 sample PWR plants. Column 4 is α values, obtained simply by dividing various PGA values by the plant SSE of 0.16g. Column 5 represents the seismic stresses corresponding to various ground motion levels, obtained as follows:

Seismic Stress = $\alpha \times$ SF $\times$ SSE value of 9.89 ksi (from Box 2 of Table 4-1)

For example, at a ground motion level of 250 cm/sec^2, α = 1.57, and Seismic Stress (at PGA of 250 cm/sec^2) = 1.57 $\times$ 0.64 $\times$ 9.89 = 9.93 ksi

In the next step, seismic stresses were combined with the normal stresses and then normalized by S_m values, (N + Seismic)/S_m, (Box 7 of Figure 4-2). These values are listed in Column 6. S_m is a material strength parameter for the pipe available from the tables in Section II of the ASME BPV Code. In Table 4-1, for Plant A, this value is 18.8 ksi. This normalization is necessary to compare these ratios with the failure criterion used in this study (Section 4.4.3), which is based on tests conducted under the EPRI Piping and Fitting Dynamic Reliability Program and subsequent evaluations performed in a separate NRC Program [Jaquay, 1998]. This failure criterion is defined in terms of multiples of S_m values. For example, for nozzles, 1% failure probability corresponds to applied stress of 4.5S_m, while 50% failure probability corresponds to a stress of 6.3S_m. In the final step, the values of (N + Seismic)/S_m versus probability of exceedance (Column 3) are plotted as shown in Figure 4-3 and compared to the failure criterion discussed above to estimate the likelihood of failure frequencies.

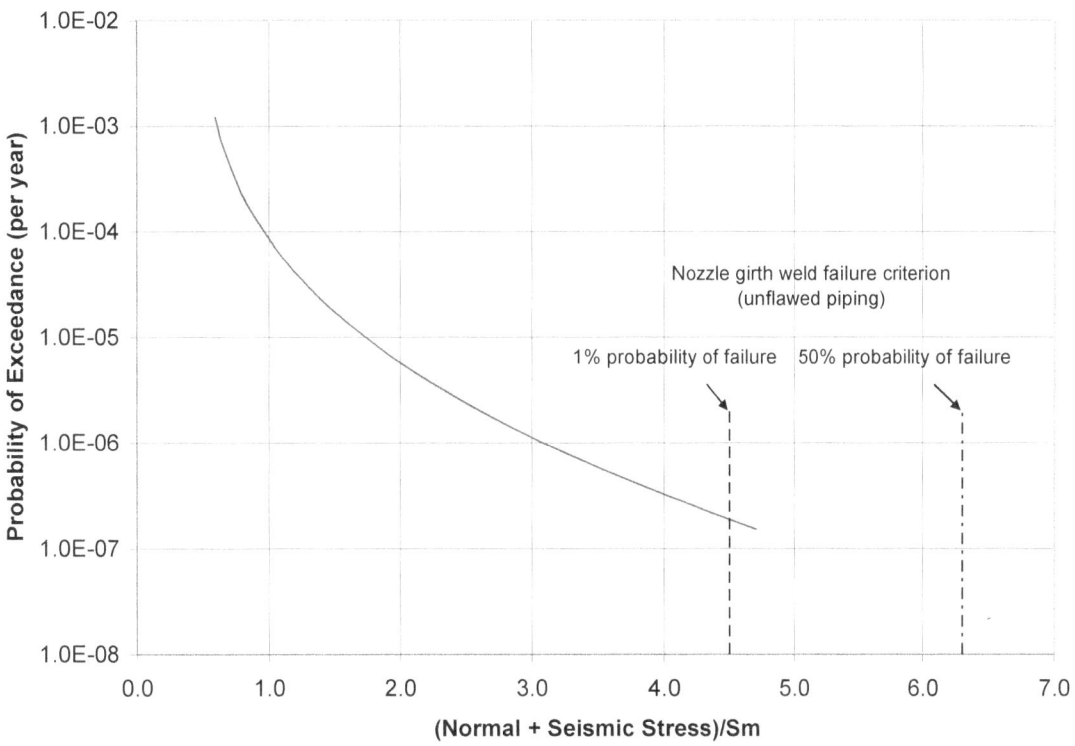

Figure 4-3   Probability of Exceedance Versus Maximum Normal + Seismic Stress in RCS Piping at a Hypothetical PWR

From Figure 4-3, it is seen that for the selected case, the probability of exceedance of stresses corresponding to 1% failure probability is approximately 10^{-7} per year. The true mean value of failure probability can be obtained by convolving mean hazard with the mean fragility function.

The following sections discuss the key steps in the above analysis in greater detail. Note that discussion related to seismic hazard, LBB database, and Scale Factors also apply to flawed piping evaluations, and the same seismic hazard values are used for indirect failure evaluations.

4.4.2 Description of the Scale Factor Approach

As discussed in connection with Boxes 4 and 5 of Figure 4-2, in the current study, it is necessary to estimate seismic stresses more realistically for various ground motion levels higher than the design basis, in order to evaluate the effects of seismic loading on the postulated TBS and the frequency used as a basis to establish the TBS. It is well documented that the seismic design methodology is often very conservative (see discussion in Section 2.1). The objective of the present study is to evaluate the likelihood of breaks for combined loading conditions of normal operating loads and seismic loads for earthquakes at annual probabilities of exceedance of 10^{-5} and lower. This objective is met by considering seismic stresses to be point estimates and about median values conditional on the earthquake considered. This means that conservatism in the seismic stresses is intended to be attributable to the use of the mean seismic hazard curve, rather than a conservative bias of the UHS or seismic analysis and design procedures. In order to remove these conservatisms, seismic fragility methods have been developed for use in seismic risk assessments and other applications. After describing some general aspects of seismic fragility in Section 4.4.2.1, Section 4.4.2.2 discusses the Factor of Safety methodology used in this study. The inverse of Factor of Safety is the Scale Factor of Box 5 of Figure 4-2. By applying this Scale Factor, seismic design loads are typically reduced by removing excessive conservatism. In the example discussed in Section 0, the value of the Scale Factor was 0.64; in other words, the SSE seismic stresses were reduced by approximately 33% to a median value conditional on an earthquake occurring with a PGA equal to the SSE and a spectral shape equal to the normalized UHS of 10^{-4} per year. Because of the importance of this factor, its application is discussed in greater detail in Section 4.4.2.3.

4.4.2.1 Seismic Fragility Methodology: General

Over the past 25 years, a seismic fragility methodology has been developed in conjunction with the development of seismic PRA techniques. The standard ANSI/ANS 58.21 [ANS, 2003] documents the requirements of the seismic fragility methodology and its application in PRA contexts. Because the current study focused only on piping system failure, a portion of ANS 58.21 [ANS, 2003] applies.

One requirement of ANS 58.21 [ANS, 2003] is for the seismic fragility evaluation to be based on a realistic seismic response that the structures, systems, and components are expected to experience at failure levels. Further, a realistic seismic response may be obtained by an appropriate combination of scaling, new analysis, and new structural models. Past seismic PRAs have embraced the following two techniques to arrive at probabilistic estimates of the fragility of soil, structures, components, piping, and equipment:

- One method is the SSMRP method, whereby the majority of risk-important structures, components, piping, and equipment are re-analyzed. Modeling, analysis procedures, and parameter values are treated as best estimates, with uncertainty explicitly introduced. These new analyses calculate seismic responses as distributions conditional on occurrence of an earthquake of a given size. Re-analysis of the vast number of systems modeled in the seismic PRA is time-consuming and resource-intensive. Practically speaking, for current applications, re-analysis is only selectively performed when design models or analysis techniques are not amenable to scaling, as discussed below. One principal purpose of the SSMRP was to help benchmark other simpler approaches.

- An alternative and simpler approach is to apply the "Factor of Safety" (FS) method, which is based on applying Scale Factors (inverse of Factors of Safety) that remove excess conservatism in the design calculated seismic response. Responses are defined as distributions in terms of a ground motion parameter (e.g., PGA). Capacities of structures, components, piping, and equipment are likewise defined as distributions in terms of the same ground motion parameter. The probability of failure can then be calculated directly for the structure, component, piping, and equipment of interest.

In the present study, the FS approach is applied to the calculated stresses in the piping systems, as determined in the SSE seismic design analyses. Application of these Scale Factors removes excess conservatisms in the SSE values introduced from a number of sources.

As previously described, the staff used point estimates in quantifying piping system failure. That is, the staff used point estimates of seismic response (which are conditional on the occurrence of various levels of ground motion) and point estimates of failure (which are conditional on stress levels in the piping system). Implementation of the complete Seismic Fragility Methodology requires treating the seismic response and fragilities as full probability distributions; this was not done for the present study, and the staff considered uncertainty through sensitivity studies.

4.4.2.2 Factors of Safety (Scale Factor is Inverse of Factor of Safety)

For the present study, the staff applied elements of the FS Method. In general, the FS Method is a technique to develop a fragility function for systems, structures, and components in terms of a ground motion parameter. The generalized form of the Factor of Safety is as follows:

$$Ac = F * Asse \qquad (4\text{-}1)$$

where Asse is a design ground motion parameter (e.g., PGA or average spectral acceleration) over a frequency range of interest; Ac is the ground motion parameter capacity; and F is the relationship between design level response and the "best estimate" of seismic capacity. Ac and F are assumed to be lognormal distributions with medians described below and logarithmic standard deviations ßc. F is expanded to represent the conservatism or Factors of Safety in the strength and response of the systems, structures, and components of interest:

$$F = Fc * Frs \qquad (4\text{-}2)$$

where Fc is the capacity factor and Frs is the response factor.

The median factor of safety, Fm, relates to the median ground acceleration capacity, Am, as follows:

$$Fm = Am / Asse \qquad (4\text{-}3)$$

The logarithmic standard deviations of F are then identical to those for the ground acceleration capacity, Ac.

The logarithmic standard deviations, ß, are further divided into a randomness, ßr, and uncertainty, ßu.

For the evaluation of a piping system failure, attributable to direct stress-induced causes, median factors of safety were generated and used to scale SSE piping stresses to best estimates at earthquake levels higher than the SSE, defined by a PGA with an annual probability of exceedance of 10^{-5}. No consideration of variability in seismic induced stresses was explicitly included (i.e., these stresses were point estimates). By contrast, for the evaluation of a piping system failure attributable to indirect causes (i.e., the result of failure of components or component supports), full fragility distributions were used.

Structure Response

Fragility of piping, components, and equipment that are supported on structures are dependent on best estimates of the seismic response of those structures. Structure response defines their seismic input. Hence, a structure response Factor of Safety, F_{rs}, is needed to remove excess conservatisms from the SSE calculated structure response.

The structure response Factor of Safety, F_{rs}, is expressed as a product of factors each representing conservatism and variability in that phase of the modeling or analysis.

$$F_{rs} = F_{ss} * F_{gmth} * F_{ssi} * F_{incoh} * F_d * F_m * F_{mc} * F_{ec} \tag{4-4}$$

where

> F_{ss} = Spectral shape factor representing the relationship between the design response spectral shape (typically SSE) and the median site-specific response spectral shape at a return period deemed important to failure; the UHS at a return period of 10,000 years was used.
>
> F_{gmth} = Ground motion time histories factor represents conservatism in the use of a single earthquake to model the design ground response spectra in the structure response analysis; this includes conservatism in enveloping of design ground response spectra by the single set of time histories and the use of a single earthquake, as opposed to an ensemble of motions more similar to actual recorded motions. This factor can also take into account the procedure of analyzing for three site soil properties and enveloping the results. [For this latter case, care must be taken to not double-count when considering soil-structure interaction (SSI).]
>
> F_{ssi} = SSI factor representing the relationship between median-centered (or best estimate) SSI modeling and the modeling used in the design analysis.
>
> F_{incoh} = Ground motion incoherency factor representing the effect of wave passage and randomness in the free-field ground motion, the effect of which is to reduce the effective foundation input motion at higher frequencies as a function of foundation size; incoherency is a significant effect for motions greater than 10 Hz.
>
> F_d = Damping factor representing the ratio of response between best-estimate damping and design damping.
>
> F_m = Factor accounting for conservatism or nonnconservatism in response, attributable to modeling assumptions.
>
> F_{mc} = Factor accounting for any bias in response attributable to modal combination rules.
>
> F_{ec} = Factor accounting for any bias in response attributable to directional combination rules.
>
> F_{nl} = Factor accounting for non-linear behavior of the structure when calculating input to supported systems (piping, equipment, components). Care must be taken to avoid triple-counting SSI effects, damping, and the non-linear response factor.

The median factor of safety and logarithmic standard deviation of structure response are as follows:

$$F_{rs} = F_{ss} * F_{gmth} * F_{ssi} * F_{incoh} * F_d * F_m * F_{mc} * F_{ec} * F_{nl} \qquad (4\text{-}5)$$

$$\beta_{rs} = (\beta_{ss}^2 + \beta_{gmth}^2 + \beta_{ssi}^2 + \beta_{incoh}^2 + \beta_d^2 + \beta_m^2 + \beta_{mc}^2 + \beta_{ec}^2 + \beta_{nl}^2)^{1/2} \ddagger \qquad (4\text{-}6)$$

Note that for structure fragility functions, the Frs factor would be multiplied by Fc, structure capacity factor, when generating the fragility function.

Piping, Equipment, and Component: Response and Capacity

This section discusses the FS approach for piping, equipment, and components in the same manner as the structure response and capacity were discussed above. In general, Eq. 4-7 applies, where A is the ground motion capacity of the piping, equipment, or component of interest:

$$A = F * A_{sse} \qquad (4\text{-}7)$$

The Factor of Safety, F, can be further divided as follows:

$$F = F_{rs} * F_{re} * F_c \qquad (4\text{-}8)$$

For structure-supported items, Frs is the term described above. Similar to the Frs factor, an Fre factor is applied when calculating best estimate component, piping, or equipment response. This factor comprises subfactors accounting for in-structure response spectral shape (e.g., peak broadening), damping of the subsystem, modeling, and various elements of analysis procedures. In addition, non-linear behavior of the component, piping, or equipment may be considered when appropriate:

$$F_{re} = F_{ess} * F_{ed} * F_{em} * F_{ean} * F_{mc} * F_{ec} * F_{enl} \qquad (4\text{-}9)$$

where

> Fess = In-structure response spectral shape factor representing conservatism in the approach taken to define the excitation applied to the subsystem; this includes conservatism in smoothing and peak broadening of in-structure response spectra. This factor accounts for design procedures, such as considering a range of soil properties, typically three, and enveloping the results. (In this latter case, care must be taken to avoid double-counting with structure response factors.)

> Fd = Damping factor representing the ratio between best estimate damping and design damping.

> Fem = Factor accounting for conservatism and nonnconservatism in subsystem modeling.

> Fean = Subsystem analysis factor to account for conservatism or nonnconservatism in the method of seismic analysis of subsystems; generally, multi-support time history analysis of multi-supported systems is considered best estimate, and other analysis procedures (such as single or multi-supported response spectrum analyses or equivalent static analyses) add conservatism to the calculated responses.

> Fmc = Factor accounting for any bias in response attributable to modal combination rules.

> Fec = Factor accounting for any bias in response attributable to directional combination rules.

---

‡ These logarithmic standard deviation values are used in the indirect failure evaluations described in Section 4.6.

Fean, Fmc, and Fec are treated in concert.

Fenl = Factor accounting for non-linear behavior of the subsystem. Care must be taken to avoid double-counting of damping and the non-linear response factor.

These Factors of Safety are also probabilistic functions described by their mean or median values and standard deviations. Most often, these functions are assumed to be lognormal distributions.

$$Fre = Fess * Fed * Fem * Fean * Fmc * Fec * Fenl \qquad (4\text{-}10)$$

$$ßre = (ßess^2 + ßed^2 + ßem^2 + ßean^2 + ßmc^2 + ßec^2 + ßenl^2)^{½} \qquad (4\text{-}11)$$

In the same manner as the structure response and equipment response factors were discussed, an equipment capacity factor is generated (Eq. 4-11). The subsystem capacity factor is often described as being comprised of a strength factor and a nonlinear ductility factor. Each factor accounts for elements of behavior between code design and failure for the failure modes of interest. The strength factor accounts for realistic material strengths and their variability. The ductility factor accounts for nonlinear behavior to failure for ductile failure modes. This capacity factor is also assumed to be lognormal.

For the case of indirectly induced piping failures, described in Section 4.6 the staff used all of the structure response factors, subsystem response factors, and subsystem capacity factors.

4.4.2.3 Application of the FS Approach

With respect to the present study, it is relatively easy to incorporate one or more of these factors to remove excess conservatisms in the calculated piping stresses. The largest sources of uncertainty and, potentially excess conservatism, in estimating the seismic stress values in piping systems arise in the definition of the seismic hazard at the site (PGA and response spectral shape), the seismic analysis methodology and associated parameters of the soil/structure system, and the dynamic response calculations of the piping system.

Table 4-3 itemizes the most important parameters to be implemented in applying the Factors of Safety in the current study. The focus is on structure response because the piping system of greatest interest is the RCL. In many cases (if not a majority of cases), it is believed that a sophisticated analysis of the RCL was performed to generate the SSE stresses used in the design and/or LBB evaluation. The intent of these analyses was to not introduce unnecessary conservatism into the results. Either a coupled model of the RCL and internal structure, or a more sophisticated uncoupled piping analysis of the RCL was performed. Hence, conservatisms associated with the actual linear piping analysis itself are judged to be minimal. Adjustment factors to account for piping system nonlinearity and ductility at high stress locations are discussed elsewhere in the report. Table 4-3 lists representative references, in which Factors of Safety are reported comparing various design and best estimate conditions. These references are representative of a large body of existing data on this subject.

Table 4-3 Important Parameters in Seismic Methodology

Structure Response Factor (Frs)	Description of the Response Factor as Applied in the Current Study
Spectral Shape Factor (Fss)	Spectral Shape Factor is the ratio of SSE spectral accelerations to the UHS spectral accelerations over a frequency range appropriate to the soil-structure-components (RCS) being considered. For soil sites, the frequency range is assumed to be 2–6 Hz. For rock sites, the frequency range is assumed to be 3–8 Hz. The frequency range can be narrowed if frequency data are available for the specific plant. NUREG-1488 [Sobel, 1994] is the basis for the UHS and SSE design response spectra.
Artificial Time History versus Ensemble (Fgmth)	Artificial acceleration time histories are generated to envelop the design ground response spectra (DGRS) (according to Section 3.7.1 of the NRC's Standard Review Plan, NUREG-0800, for example). These histories introduce conservatisms beyond the stated goal of a mean-plus-one-standard-deviation response conditional on the PGA of the DGRS. This conservatism has been quantified in various forms by Maslenikov et al. [995].
Structure Damping (Fd)	Structure damping is an important parameter especially for rock sites. (For soil sites, SSI typically dominates the important parameters to structure response.) There are several different techniques to account for changes in damping. For cases of broad-banded ground response spectra and a general lack of detailed specific information, the Scale Factors or Factors of Safety are best described by ratios developed using ground response spectra from NUREG/CR-0098 [Newmark and Hall, 1978]. For purposes of this study, lacking detailed dynamic response information, the frequency range of interest is assumed to be 2–8 Hz. The best-estimate structure damping value is that damping corresponding to structure behavior at the earthquake ground motion level corresponding to the failure level of piping or components. For reinforced concrete, this damping level is likely to be in the range of 7–10%.
SSI Kinematic Interaction — including Incoherence (Fssi & Fincoh)	SSI is a most important phenomenon when considering soil sites. Two aspects of SSI are kinematic interaction and inertial interaction. Kinematic interaction is the effect of spatial variation in ground motion on the "foundation input motion" to the structure. Two kinematic interaction phenomena are (1) spatial variation of ground motion with depth in the soil (i.e., from soil-free surface through the depth of soil to the bottom of the embedded foundation), and (2) the effect of ground motion incoherence over a horizontal plane. The spatial variation of ground motion with soil depth can be important in two regards. Specifically, (1) older plants were designed to DGRS or time histories applied directly to the foundations of structures without accounting for any ground motion variation with soil depth, and (2) in the late 1970s and early 1980s, the requirement was to place the DGRS at foundation level. Both cases add conservatism to the calculation of SSI and structure response. Inertial interaction is the combined dynamic response behavior of the soil-structure system. Incoherence denotes randomness as a function of frequency and distance between observation points. Higher frequencies (greater than 10 Hz) and distances comparable to nuclear power plant foundations effectively filter this higher-frequency motion. For this study, the incoherency factor was not applied.

Structure Response Factor (Frs)	Description of the Response Factor as Applied in the Current Study
SSI Inertial Interaction (Fssi)	Inertial interaction, the combined dynamic response behavior of the soil-structure system, is especially important for nuclear power plant types of structures and foundations (large, stiff, mat foundations). "Equivalent damping" for inertial interaction can range from relatively low values for rocking to very large values (greater than 50%) for horizontal and vertical deformations. Quantification of the effects of SSI on structure response has been studied by Maslenikov et al. [1995] and by Johnson and Maslenikov [1984].
Structure Model (Fm)	Conservatism or nonconservatism in the structure dynamic model leading to over- or under-prediction of dynamic response.

Seismic Hazard and Spectral Shape

A key element of the current study is the seismic hazard reported in NUREG-1488 [Sobel, 1994], which contains 1993 LLNL seismic hazard curves (Appendix A) for each of 69 nuclear power plant sites in terms of PGA and UHS for 5% damping (Appendix B) for numerous return periods. In both cases, mean, median, and 15% and 85% non-exceedance probabilities are presented. The mean seismic hazard curves were used in this evaluation. The median UHS at a return period of 10,000 years, normalized by peak acceleration was used to define the best-estimate spectral shape. Note that the UHS for other lower annual probabilities of exceedance (10^{-5} or 10^{-6}) was not available from NUREG-1488 [Sobel, 1994]. However, because the normalized UHS spectral shape was used in the calculation, the normalized 10^{-4} per year UHS was deemed to be representative of the lower probability UHS. Appendix C [Sobel, 1994] also provides the SSE data for the majority of units to be considered. Note that the seismic hazard (seismic hazard curves and UHS) is for the site; generally, the SSE design ground motion may differ unit-by-unit for a particular site due to the vintage of the unit, the state of knowledge of earthquake engineering, and the NRC regulations.

The Factor of Safety for spectral shape was developed by applying the following steps:

- Normalize the SSE design ground response spectra (5% damping) to a PGA of 1.0 at 33 Hz.

- Calculate the PGA corresponding to the mean hazard at a return period of 10,000 years. Anchor the 10,000-year return period UHS (5% damping) to this PGA at 50 Hz.

- Calculate an average spectral acceleration over the frequency range of most interest to the structures housing and supporting the NSSS:

    o   For rock-founded sites, this frequency range is approximately 3–8 Hz. The high end of this frequency range is most applicable to the combined internal structure/RCL. The lower frequency range includes the dynamic behavior of the reactor containment structure. Both are intended to encompass dynamic behavior of uncracked and cracked sections because the excitation level of interest is at annual probabilities of exceedance of 10^{-4} or less. A lack of specific frequency information about the plants of interest prevented narrowing this range; however, for the majority of rock-founded plants, the staff only performed a sensitivity study on Factors of Safety accounting for spectral shape. For the frequency range of 3–8 Hz, the average Factor of

Safety is 1.31 with a range from about 0.7 to 2.2 (i.e., on average, the SSE spectral shape in the frequency range of 3–8 Hz is conservative by about 30%). For an expanded frequency range of 3–12 Hz, the average Factor of Safety is 1.21 with a range of about 0.65 to 1.9. For the frequency range of 8–12 Hz, the average Factor of Safety is 1.065 with a range of about 0.6 to 1.7. Generally, these values emphasize that the UHS for rock sites has greater frequency content in the higher frequency range (i.e., greater than 10 Hz). Quantitatively, this fact has the stated effect on the Factor of Safety. Note that this is not the complete story of conservatism or nonconservatism, because other seismic analysis parameters and procedures enter the evaluation. (Such other parameters and procedures include structure and subsystem damping, analysis techniques, and SSI kinematic interaction phenomena.)

- For soil-founded structures, the frequency range is assumed to be 2–6 Hz. Soil foundations significantly reduce the frequencies of the system for low excitation levels and even further reduce these frequencies at higher excitation levels because of the softening of soil stiffness with increased strains in the soil.

Implementation was done in two stages. First, the staff selected three plants for initial evaluation — one CE plant on a soil site, one Westinghouse plant on a soil site, and one Westinghouse plant on a rock site. Second, the staff evaluated all PWR plants sited at rock sites, where design stress information was available. The following discussion presents the results of initial studies for illustrative purposes.

Combustion Engineering (CE) Plant

The selected CE plant is sited on a deep soil site. To determine the Factors of Safety, the staff reviewed all available information and derived Factors of Safety. This case demonstrates that one of the potentially most significant factors in determining best estimate response for plants founded on soil is the Factor of Safety related to soil-structure interaction, especially for plants designed to NRC regulations prior to the mid-1980s. Table 4-4 shows the elements of the structure response factor.

Table 4-4 Factors of Safety: Structure Response for CE Plant

Response Factor	Basis for Response Factor	Median Factor
Spectral shape (Fss)	Spectral shape response factor based on comparison of NUREG-1488 [Sobel, 1994] UHS and SSE over frequency range 2–6 Hz	0.86
Ground motion time histories (Fgmth)	Assumed to be included in Fssi	1.0
Soil-structure interaction (Fssi)	UCID-20122, Vol 1 [Johnson and Maslenikov, 1984], Table 4.6, Average of Cases 17/3, 18/4, 17/7, and 18/8, to approximate Vs=1500fps — embedded foundation/surface foundation — all analyses were best estimate; likely conservative, not enough information to refine	2.42
Incoherence (Fincoh)	Incoherence is an effect for frequencies greater than 10 Hz	1.0
Damping (Fd)	Damping SSE (5%) vs. UHS (7%) — factor based on ratio of damping for median NUREG/CR-0098 response spectra [Newmark and Hall, 1978]	1.12
Modeling (Fm)	Assume best estimate	1.0
Modal combination (Fmc)	Time history analysis — modal combination by algebraic sum	1.0
Directional combination (Fec)	Time history analysis — directional combination by algebraic sum	1.0
Nonlinear (Fnl)	Assume nonlinear effects included in damping factor	1.0
Total (Frs)		2.33

Two interpretations of the structure response Factor of Safety are as follows:

- For fragility analyses of the reactor building and its contents, the SSE PGA is scaled by 2.33 to account for conservatisms in the seismic structure response calculations. In addition, subsystem response factors of safety would apply, as would capacity factors for the structure or subsystem of interest.

- For direct stress-induced failure evaluation of the RCL piping, a Scale Factor equal to the inverse of the Factor of Safety is to be applied to the stress scaled to higher-level earthquake loadings. The Scale Factor is (1/2.33) = 0.43.

Westinghouse Plant (Soil Site)

The selected Westinghouse unit is sited on a soil site of thickness in the range of 80–180 ft. over bedrock. To determine the Factors of Safety, the staff reviewed all available information and derived Factors of Safety. Table 4-5 shows the elements of the structure response factor.

Table 4-5 Factors of Safety: Structure Response for Westinghouse Plant — Soil Site

Response Factor	Basis for Response Factor	Median Factor
Spectral shape (F_{ss})	Spectral shape response factor based on comparison of NUREG-1488 [Sobel, 1994] UHS and SSE over frequency range 2–6 Hz	2.11
Ground motion time histories (F_{gmth})	Assumed to be included in F_{ssi}	1.0
Soil-structure interaction (F_{ssi})	UCID-20122, Vol 1 [Johnson and Maslenikov, 1984], Table 4.2, Case 15/13 — embedded foundation/surface foundation/shallow soil layer/Vs=1000fps.	1.74
Incoherence (F_{incoh})	Incoherence is an effect for frequencies greater than 10 Hz	1.0
Damping (F_d)	Damping SSE (7%) vs. UHS (10%) — Damping values based on NUREG/CR-0098 [Newmark and Hall, 1978]	1.15
Modeling (F_m)	Assume best estimate	1.0
Modal combination (F_{mc})	Time history analysis — modal combination by algebraic sum	1.0
Directional combination (F_{ec})	Time history analysis — directional combination by algebraic sum	1.0
Nonlinear (F_{nl})	Assume nonlinear effects included in damping factor	1.0
Total (F_{rs})		4.22

Two interpretations of the structure response factor of safety are as follows:

- For fragility analyses of the reactor building and its contents, the SSE PGA is scaled by 4.22 to account for conservatism in the seismic structure response calculations. In addition, subsystem response factors of safety would apply, as would capacity factors for the structure or subsystem of interest.

- For direct stress-induced failure evaluation of the RCL piping, a Scale Factor equal to the inverse of the Factor of Safety is to be applied to the stress scaled to higher-level earthquake loadings. The Scale Factor is (1/4.22) = 0.24.

Westinghouse Plant (Rock Site)

The selected Westinghouse unit is sited on a rock site. To determine the Factors of Safety, the staff reviewed all available information and derived Factors of Safety. Table 4-6 shows the elements of the structure response factor.

Table 4-6   Factors of Safety: Structure Response for Westinghouse Plant — Rock Site

Response Factor	Basis for Response Factor	Median Factor
Spectral shape (F_{ss})	Spectral shape response factor based on comparison of NUREG-1488 [Sobel, 1994] UHS and SSE over frequency range 3–8 Hz.	1.60
Ground motion time histories (F_{gmth})	Assumed to be best estimate — probably conservative	1.0
Soil-structure interaction (F_{ssi})	Rock site; no kinematic interaction effect.	1.0
Incoherence (F_{incoh})	Incoherence is an effect for frequencies greater than 10 Hz	1.0
Damping (F_d)	Damping SSE (5%) vs. UHS (10%) — Damping values based on NUREG/CR-0098 [Newmark and Hall, 1978]	1.29
Modeling (F_m)	Assumed to be best estimate	1.0
Modal combination (F_{mc})	Time history analysis — modal combination by algebraic sum	1.0
Directional combination (F_{ec})	Time history analysis — directional combination by algebraic sum	1.0
Nonlinear (F_{nl})	Assume nonlinear effects included in damping factor	1.0
Total (F_{rs})		2.06

Two interpretations of the structure response factor of safety are as follows:

- For fragility analyses of the reactor building and its contents, the SSE PGA is scaled by 2.06 to account for conservatism in the seismic structure response calculations. In addition, subsystem response factors of safety would apply, as would capacity factors for the structure or subsystem of interest.

- For direct stress-induced failure evaluation of the RCL piping, a Scale Factor equal to the inverse of the Factor of Safety is to be applied to the stress scaled to higher-level earthquake loadings. The Scale Factor is (1/2.06) = 0.48.

The staff repeated the above process for all selected plants and developed Scale Factors to estimate seismic stresses at various ground motion levels.

4.4.3 Brief Description of Unflawed Piping Failure Criterion

Section 2.1 defined unflawed piping as a piping system that may have such small flaws that it essentially behaves as having its full cross-sectional area as the load resisting area. Under cyclical loads, failures in such piping systems are primarily caused by fatigue ratcheting, in which a crack or flaw initiates and propagates during the application of the cyclical loads. One source of applicable test data for such behavior is EPRI's Piping and Fitting Dynamic Reliability Program [EPRI, 1995] and the subsequent evaluation performed in a separate NRC Program [Jaquay, 1998]. Recent revisions to seismic design rules in the 2004 ASME BPV Code revisions are based on these studies.

The EPRI program tested some 37 piping components under dynamic loading, and the NRC's evaluation of these tests [Jaquay, 1998] developed a fragility function to characterize the probability of failure versus given elastic stress levels. This function is based on applying safety factors to the allowable values specified by the ASME BPV Code. In general, a safety factor of 2 was applied to stresses corresponding to 1% probability of failure. For straight-pipe girth welds, the Service Level D limits of $3S_m$ imply that there is 1% probability of failure at stress values of $6S_m$. Table 4-7 summarizes the stress values corresponding to 1% failure probability for components of interest.

Table 4-7 Stress Values at 1% Failure Probability

Component of Interest	Stress Values Corresponding to 1% Failure Probability
Straight-pipe girth weld	$6 S_m$
Nozzle girth-weld	$4.5 S_m$
PWR Elbows	$4.2 S_m$ to $6 S_m$
BWR Elbows	$2.9 S_m$ to $4.1 S_m$

Note that elbow values are variable depending on bend radius, mean radius, and thickness of the pipe system, which differs considerably for PWR and BWR plants.

Stress values corresponding to 50% failure probabilities are 1.4 time the stress values corresponding to 1% failure probability for straight pipe and elbows. For nozzle welds, the stress value corresponding to 50% failure probability is $6.3 S_m$.

For PWR plants, the bounding failure criterion is probably at the nozzle girth welds, given that reviews of piping stresses in LBB submittals suggest that these are typically the highest stress locations. Occasionally, there is a high, unintensified stress at an elbow, but the lower bound of the elbow failure criterion is close to that of the nozzle girth weld.

For BWR plants, the bounding failure criterion is probably at the elbows, given that these are thinner-walled pipes with much higher stress intensity indices.

It is also important to note that the above values are derived from the EPRI test program, and were based on tests with piping diameters of 3 to 6 inches (7.6 to 15.2 cm). In the present study, the staff assumed that these new design rules apply equally to piping of all diameters; hence, pipe size does not alter this unflawed pipe seismic failure criterion.

4.4.4 Key Results and Findings

The steps described in Section 4.4.1 and illustrated by an example were repeated for each of the 27 PWR plants selected for the present study. As stated, the staff used mean LLNL seismic hazard curves corresponding to each selected site, and developed Scale Factors using the approach described in Section 4.4.2.2 with the normal and SSE stresses and materials listed in the LBB database. In virtually all LBB submittals, the highest stress location is at (or very close to) a nozzle, so the failure criterion for the nozzle girth welds is most applicable. Figure 4-4 shows results from the 27 PWRs for the most highly stressed hot leg, cold leg, or cross-over (suction) legs. As seen from these plots, and as expected, the unflawed piping has very low probability of failure attributable to seismic loads. The earthquake experience of piping systems has also repeatedly demonstrated good performance (see Section 2.3).

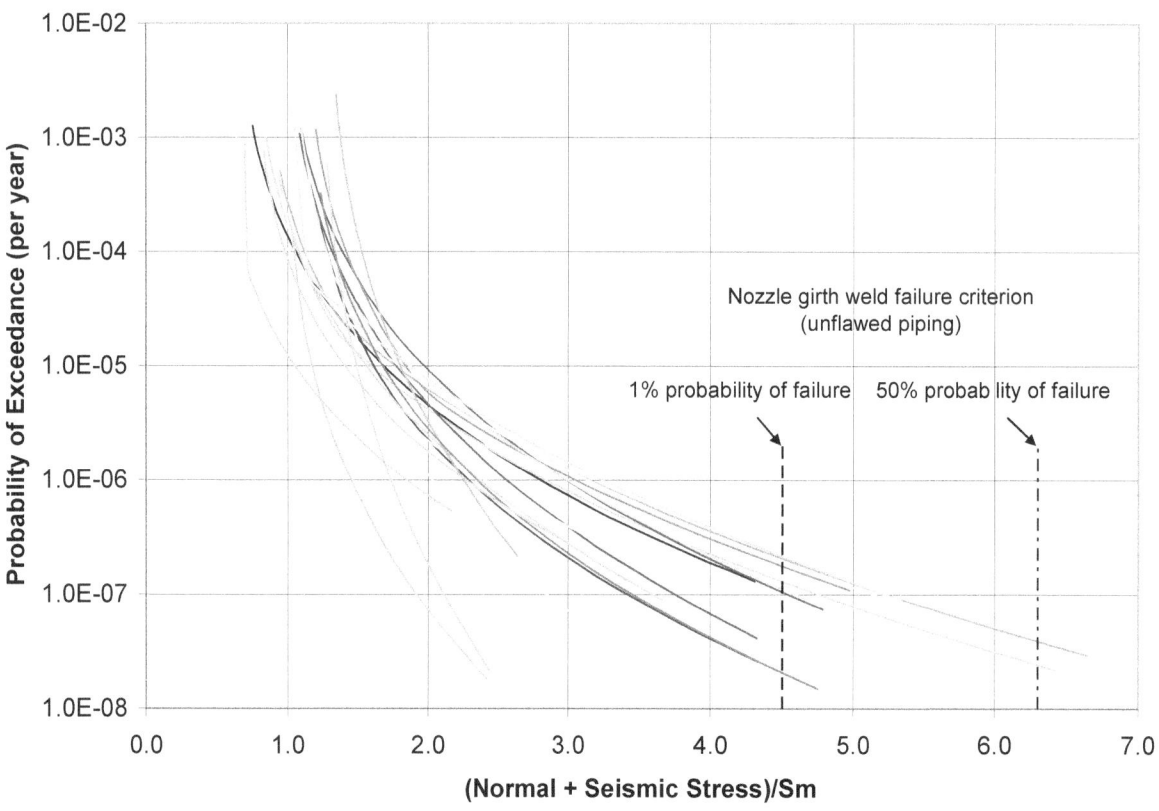

Figure 4-4 Probability of Exceedance Versus Maximum Normal + Seismic Stress in RCS Piping at 27 PWRs

Figure 4-5 shows results for pressurizer surge lines in six PWRs, with the same overall conclusions. Based on these findings and results of other studies, the staff concludes that unflawed PWR piping has failure frequencies much less than 1×10^{-5} per year.

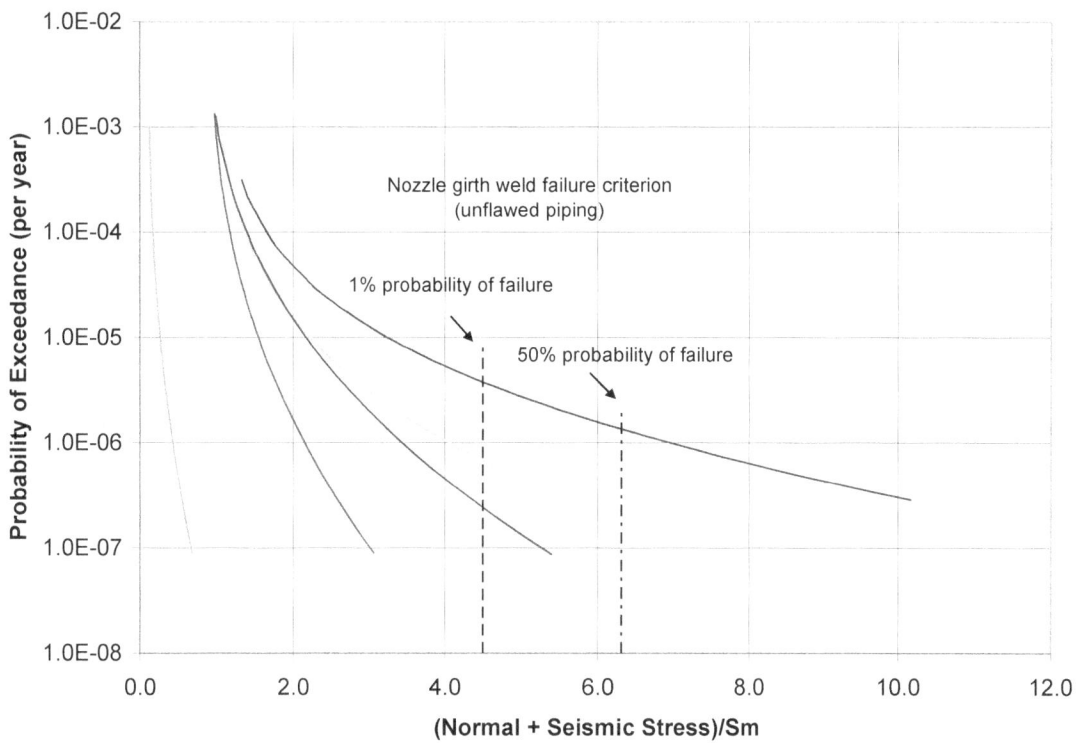

Figure 4-5 Probability of Exceedance Versus Maximum Normal + Seismic Stress in Pressurizer Surge Piping at Six PWRs

4.5 Flawed Piping

4.5.1 Approach, Assumptions, and Key Steps

As stated in Section 2.2, for piping systems with flaws, the staff based its approach on the "flaw exclusion" principle rather than the detailed probabilistic fracture mechanics approach discussed in Section 3. Two approaches are presented in this section. One examines critical surface cracks at the 10^{-5} and 10^{-6} seismic stresses, while the other approach is similar to a LBB analysis, but for critical circumferential through-wall cracks at the 10^{-5} and 10^{-6} seismic stresses.

The objective of the first approach is to estimate the surface flaw size (depths and lengths) of circumferentially oriented surface flaws in primary system piping that would be predicted to be "critical" (i.e., would lead to piping failure) if they were subjected to severe earthquake loading, and to compare these critical surface flaw size estimates with the "allowable" surface flaw size estimates according to ASME Section XI requirements. It should be noted, however, that the outcome of both surface flaw analyses is a locus of surface-crack geometries [(i.e., graphs relating the surface flaw depth-to-thickness ratios (a/t) to the circumferential surface crack lengths normalized by the pipe circumference (θ/π)], that are predicted to be allowable (or critical).

Specifically, the approach for evaluation of surface-flawed piping is to determine whether the maximum allowable flaw sizes from an ASME Section XI flaw-evaluation procedure using the past design N+SSE stresses (Service Level D Code evaluation with all the Code safety factors and conservatisms) gives surface flaw sizes that are smaller than critical flaw sizes for 10^{-5} annual probability of exceedance seismic events using the analysis methodology described herein.

As discussed in Section 4, the staff conducted this evaluation for a few selected piping systems of PWR plants that are located east of the Rocky Mountains. This analysis primarily considered pipe sizes larger than the TBS. For PWRs, the large-diameter primary piping systems included hot leg, cold leg, and crossover leg piping, and the staff also performed an analysis of a surge line.

The cases chosen were for plants with wrought stainless steel primary pipe systems (predominantly Westinghouse plants) and plants with carbon steel piping (predominantly CE plants). The normal operating and SSE stresses for a large number of PWR plants were available from a database created from past LBB submittals (Section 4.2). A Westinghouse piping system fabricated from cast stainless steel was also analyzed. The analysis methodology developed here should be appropriate for this material, provided that the fracture toughness properties have not significantly degraded as a result of thermal aging.

The flow chart in Figure 4-6 gives an overview of the process for determining Code allowable surface flaw sizes at design stresses and critical flaw sizes for N+10^{-5} [§] stresses. The same procedure is used to determine critical surface flaw sizes for N+10^{-6} stresses. As shown in the figure, the process begins with selection of a piping system from the LBB database, and obtaining N+SSE design stresses at critical locations of interest. The subsequent steps are detailed in the following subsections (as indicated in the flow chart) followed by an example case in Section 4.5.2.9, and a discussion of overall results in Section 4.5.2.10.

The second approach is similar to a LBB analysis, but for critical circumferential through-wall cracks at the 10^{-5} and 10^{-6} seismic stresses. In this case, the leak-rate is calculated based on the normal operating stresses and the assumptions on the type of cracking mechanism (i.e., air fatigue, corrosion fatigue, or PWSCC). The LBB approach in the NRC's draft Standard Review Plan (SRP) 3.6.3 has a requirement that the pipe system of interest is not susceptible to mechanisms that will result in long surface cracks, i.e., creep, erosion-corrosion, and stress corrosion cracking. SRP 3.6.3 also uses the plant technical specifications leakage requirement (1 gpm for PWRs and 5 gpm for BWRs) with a safety factor of 10 on the leak rate. In a few later LBB submittals, the plants were able to justify using 0.5-gpm leakage detection for their LBB analysis. The leakage size flaw (with the safety factor on leakage) is then doubled and compared to the critical flaw size at the N+SSE stresses in the SRP 3.6.3 approach. A similar approach is used for the 10^{-5} and 10^{-6} seismic stress ratios with and without safety factor on the critical flaw size. This was done using the same procedures for modification of the seismic stresses as used in the surface flaw evaluation procedures, and for a subset of the same PWR plants east of the Rocky Mountains.

[§] In this section, the shorthand notation "N+10^{-5} stresses" or "N+10^{-6} stresses" is used for convenience to describe combined normal operational stresses and seismic stresses attributable to earthquake ground motion levels associated with the 10^{-5} or 10^{-6} annual probability of exceedance, respectively.

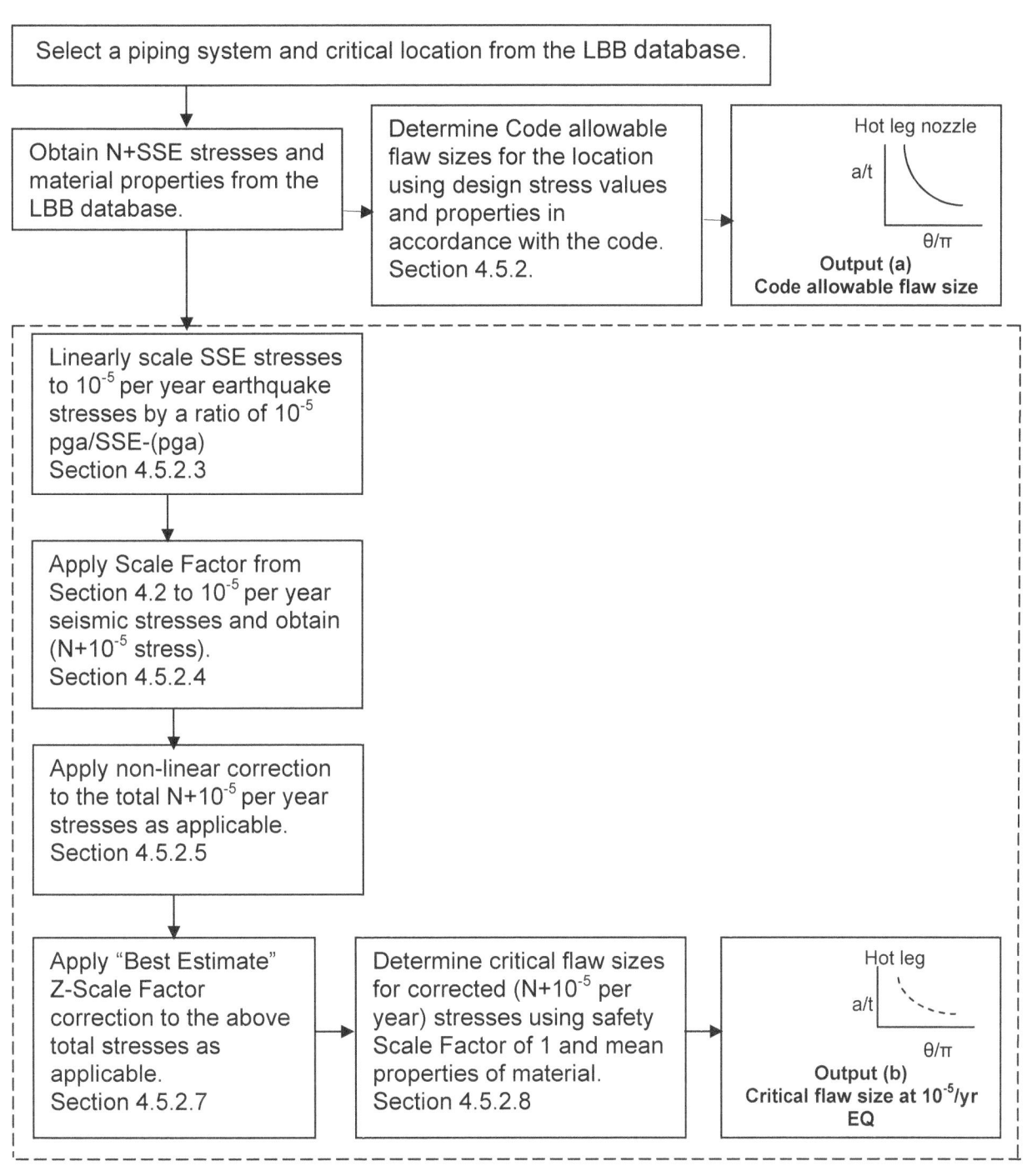

Figure 4-6 Critical Flaw Size Determination Procedure

4.5.2 Surface Flaw Evaluation Procedures

The ASME maximum allowable surface flaw size procedures at N+SSE stresses, and the critical surface flaw evaluation procedures at 10^{-5} and 10^{-6} stresses are given below.

4.5.2.1 Details of the ASME Surface Flaw Evaluation Procedure

The ASME flaw evaluation procedure used in this study is detailed in the 2004 version of Appendix C to Section XI. This version includes evaluation procedures for both austenitic and for ferritic pipes. The types of materials that can be evaluated include ferritic welds, ferritic base metals, wrought stainless steels, stainless steel tungsten inert gas / metal inert gas (TIG/MIG) welds, stainless steel submerged arc weld/shielded metal arc weld (SAW/SMAW) welds, and cast stainless steel base metals with less than 20% ferrite, (i.e., cast steels that are not susceptible to significant thermal aging). Code procedures do not address Inconel 82/182 welds or other cast stainless steels that might be more susceptible to thermal aging.

The Code flaw evaluation procedures are based on a limit-load analysis for a circumferential flaw, where the allowable or critical flaw size is estimated as a function of the applied loads and material strength at the temperature and loading rate of interest. This type of analysis is only valid for materials having sufficient fracture toughness so that their limit load is controlled only by their flow strength. Such high-toughness materials include wrought stainless steels and stainless steel TIG/MIG welds. Other materials have lower limit loads that are controlled by their elastic-plastic fracture toughness (J-R curve) rather than their flow strength. To account for the lower limit loads of these materials the Code uses a multiplier on the applied stress called the "Z-factor." The Z-factor is the ratio of the failure stress predicted on the basis of flow strength to the failure stress predicted using an elastic-plastic fracture mechanics analysis and the J-R curve for the piping material of interest. The Z-factor is a function of the material and the pipe diameter. The Code gives Z-factors for certain materials, but does not address all locations where flaws could exist in the piping systems of interest. The allowable flaw size evaluations performed using Code procedures were only conducted for those materials for which Z-factors exist. The critical flaw size evaluations for N+10^{-5} stresses included estimates of case-specific Z-factors for the various materials and piping systems of interest.

In the ASME Code evaluation procedure, the total applied stress is partitioned into the following components:

- primary membrane stress (P_m)
- primary bending stresses (P_b)
- and secondary bending stresses (P_e) which include the effects of thermal expansion and SAM

The Safety Factors (now called Structural Factors in the 2004 version of the Code) prescribed by the Code are different for each of these stress components. Past full-scale tests on lower-toughness welds with circumferential cracks [Wilkowski et al., 1998] revealed that weld residual stresses do not contribute to elastic-plastic fracture for these materials; hence, such stresses are not included in either the Code or the best-estimate flaw evaluation analyses described herein[**]. The SFs also change depending on service level. For seismic loading, Service Level D

[**] Residual stresses could affect the failure stress of cast stainless steel if very severely affected by thermal aging.

conditions are used, and the corresponding SF is 1.3 for membrane stress, 1.4 for bending stress, and 1.0 for thermal expansion stress.

The LBB data, discussed in Section 4.2, provide estimates of the forces and moments on the piping based on elastic stress analysis. The unintensified stresses are then calculated for the piping using the pipe dimensions and other information. The plant piping forces and moments are typically given for normal operating conditions and normal plus SSE conditions. The normal operating values include pressure, dead weight, and thermal expansion (all combined together). Because the pressure is known, the membrane stress (P_m) included only the pressure-induced axial stress. The rest of the normal operating stresses were taken as primary bending stresses (P_b) in this initial analysis, given that the dead weight, thermal expansion, and any bending stresses from pressure are not separated. The SSE stresses were calculated from the bending moments given in the LBB reports, and were assumed to be all inertial or primary bending stresses (P_b). A sensitivity study was conducted to evaluate effects if some part of the normal operating stresses was assumed to be thermal expansion stresses with a lower SF in the ASME flaw size evaluation. Based on this study, stresses were categorized into primary stresses and secondary stresses in the Code analyses. This causes the Code-allowable flaw to be larger than those based on the assumption of all primary stresses.

In the Code analyses, the material yield and ultimate strengths were taken from Section II of the Code at the operating temperature. These values are used to calculate the "flow stress" of the material in the Code limit-load analysis procedure. The Code defines the flow stress to be the average of the yield and ultimate strengths. Actual strength properties are typically higher than the Code values, so there is an inherent degree of conservatism in these calculations.

The Code analysis procedures used are viewed as producing conservative estimates of allowable flaw size because of the following factors:

(1) applied SFs

(2) use of Code material strengths

(3) use of elastic stresses if the total stresses are above yield

(4) the significant conservatism of the ferritic material Z-factors, while the austenitic SAW/SMAW Z-factors are reasonable when compared to full-scale pipe fracture experimental data

With regard to item (2) above, the Code allows the use of actual material strength values if the Certified Material Test Reports (CMTR) are available. Both Code material strengths and CMTR values are used in the results reported herein. The use of CMTR properties results in estimates of appreciably larger allowable flaw sizes for some materials. These conservatisms are relaxed in the procedure used to estimate critical flaw sizes for $N+10^{-5}$ stresses discussed in Section 4.5.2.2.

4.5.2.2 Details of Critical Surface Flaw Sizes for $N+10^{-5}$ and $N+10^{-6}$ Stresses

The evaluation of the critical flaw sizes associated with $N+10^{-5}$ and $N+10^{-6}$ stresses involves the following steps:

(1) Scale the SSE stresses to the stresses for the 10^{-5} or 10^{-6} annual probability of exceedance event.

(2) Apply seismic scale factors as presented in Section 4.4.2.

(3) Account for the fact that the operational and seismic stresses are from elastic analyses, and actual stresses would be lower due to the nonlinear behavior of the material.

(4) Use estimated actual strengths for the materials being evaluated.

(5) Use Z-factors (elastic-plastic fracture correction to limit-load analyses) that were recalculated for each material of interest where cracks could occur.

After completing these steps, the staff calculated the critical flaw sizes for $N+10^{-5}$ or $N+10^{-6}$ stresses for each case. The following sections present some additional details for each step.

4.5.2.3 Scaling of SSE Stresses to 10^{-5} or 10^{-6} Annual Probability of Exceedance Seismic Event

For each case selected for analysis, the maximum normal and SSE stresses are known at some location along the length of the pipe system from information compiled in the LBB database. This maximum combination of N and SSE stresses occurs at a single location, and does not represent the combination of highest individual N and SSE stresses at locations along the pipe segment. Typically this location of maximum stresses occurs close to a nozzle where there may be a safe end, several types of welds, and base metals.

As discussed in Section 4.3, the staff used the LLNL hazard curves in this study. These curves are given in tabular form for the mean value of PGA versus annual probability of exceedance. The PGA associated with the SSE is also known from plant-specific information. From this information, the staff determined the accelerations that corresponded to the 10^{-5} or 10^{-6} annual probability of exceedance seismic event. The staff then linearly scaled the seismic stresses for the 10^{-5} or 10^{-6} annual probability of exceedance seismic event by setting them equal to the SSE stress times the ratio of the PGA for the 10^{-5} or 10^{-6} annual probability of exceedance seismic event to the PGA for the SSE.

4.5.2.4 Scale Factors to Modify Seismic Stresses

As discussed in Section 0, the staff developed a Scale Factor to account for the (1) shape differences between UHS and design spectra used, (2) differences between the damping of building structures from design and best-estimate values, and (3) differences in soil-structure interaction analysis methods between design and best-estimate values. The staff then multiplied the linearly scaled seismic stresses by this Scale Factor to obtain a more realistic estimate of the seismically induced inertial bending stress. Finally, the staff added the normal (N) stresses to this value to determine the total stresses.

4.5.2.5 Nonlinear Correction to Linearly Scaled Stresses

Although the Code Section III stress analysis and Section XI pipe flaw evaluation procedures use elastic stresses, experimental and analytical evaluations have shown that a pipe in bending will exhibit considerable nonlinear behavior even when it contains circumferential flaws of significant size. Moreover, significant conservatism can be associated with the use of elastic stresses in nonlinear fracture mechanics, especially for pipes containing flaws that are small enough that the unflawed pipe in the system would have stresses above the yield strength of the material. This leads to substantial discontinuity between unflawed pipe failure criterion developed for the new seismic rules and the critical flaw sizes for the 10^{-5} or 10^{-6} annual probability of exceedance seismic events that are evaluated here. As a result, a nonlinear correction to the elastic stresses was approximated. The basis for the approximate solution is provided in the following paragraphs.

The failure criterion for unflawed pipe was examined in the EPRI Piping and Fitting Dynamic Reliability Program [EPRI, 1995] as discussed in Section 4.4.3. That work involved unflawed pipe and fitting tests that were shaken to failure. The failure mechanism was typically ratchet fatigue. These tests were the basis for the changes to the new ASME Section III seismic design rules. In those rules, the maximum allowable seismic stress is $3S_m$ for girth butt welds[††]. This limit of $3S_m$ is placed on pressure, dead-weight stresses, and seismic inertial stresses, but does not include the stresses produced by thermal expansion or by SAM. The Section III seismic design stress of $3S_m$ is an elastically calculated value and includes a safety factor of 2. Consequently, the value of maximum allowable seismic stress is actually $6S_m$, based on the work reported in Appendix III-B to NUREG/CR-5361 [Jaquay, 1998]. Furthermore, the data in NUREG/CR-5361 [Jaquay, 1998] show that $6S_m$ corresponds to a 1% probability of failure. The mean probability of failure is a factor of 1.4 higher, or $8.4S_m$, for girth butt welds. For nozzles, the mean failure stress is $6.3S_m$. However, for static loading with nonlinear behavior, the unflawed thick-walled pipe maximum moments have an elastically estimated bending stress approximately equal to the flow stress, which is equal to the average of the yield and ultimate strengths, $[0.5(S_y+S_u)]$. These results provide the basis for a first-order correction of the elastic stresses reported in the LBB database, as shown in Figure 4-7. If the elastic stresses are below the yield strength of the material the correction factor is 1.0. If the elastic stresses exceed the yield strength of the material then the correction factor diminishes from unity to the ratio of $0.5(Sy+Su)/6.3S_m$. The correction presented in Figure 4-7 was used in the critical flaw size analyses reported herein.

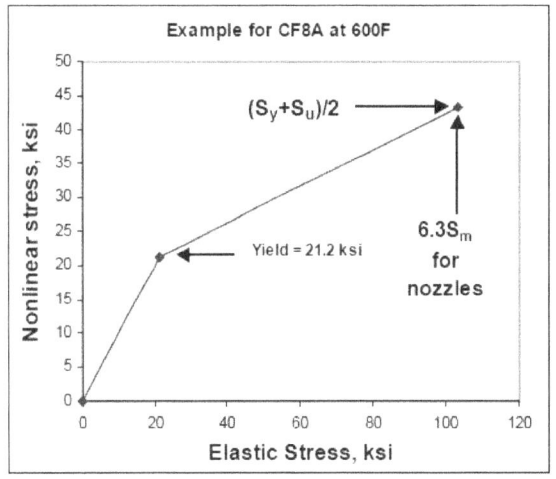
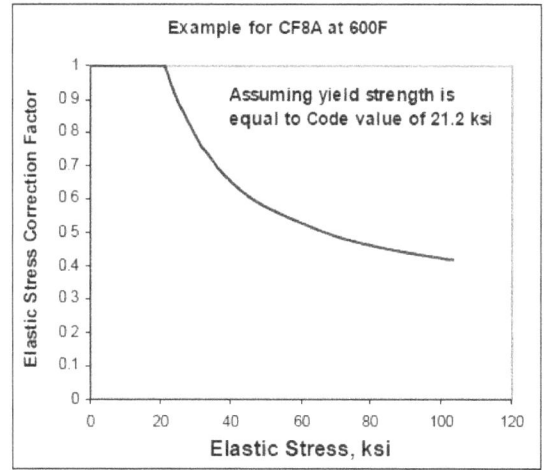

Figure 4-7 **Simple Elastic-Stress Correction Curves for Typical Nuclear Piping Steels Using Code Strength Values** (Note that a lower correction would exist using actual yield strength values, which are typically higher than Code values.)

[††] The elastic stress correction Scale Factor is, therefore, based on failure of unflawed pipe butt welds, although a lower value would exist for butt welds close to nozzles or in the thinner BWR elbows. It is a relatively easy matter to revise the calculations in this report to use the unflawed nozzle girth weld stresses instead. That would give a smaller elastic-stress correction factor.

4.5.2.6 Actual Strength Values

The staff used the actual strength values in the critical flaw evaluations for CF8A for austenitic piping and A516 Grade 70 for ferritic piping. For the CF8A base material, the mean strength values were based on data presented in NUREG/CR-6004 [Rahman, 1995a]. The mean was based on 45 stress-strain curves tested at temperatures from 550° to 608 °F for different cast stainless steels having different amounts of thermal aging. For the CF8A the "mean average" yield strength was 38% higher than the Code value at 600 °F and the ultimate strength was 18% higher than the Code value at 600 °F.

For the A516 Grade 70 ferritic pipe steel, the mean value used was an average based on three tensile specimens removed from different locations through the thickness of a CE cold-leg procured during the NRC's Degraded Piping Program [Wilkowski et al., 1989]. These tests were performed at 550 °F The "mean average" yield strength was 23% higher than the Code value at 550 °F, although the ultimate strength was only 2% higher than the Code value at 550 °F.

4.5.2.7 Best-Estimate Z-Factors

As noted in the discussion of the allowable flaw size estimation (Section 4.5.2), the Z-factor is a simple way of accounting for the effects of lower fracture toughness on the failure stress. The Z-factor is the ratio of the failure stress predicted on the basis of flow strength to the failure stress predicted using an elastic-plastic fracture mechanics analysis and the J-R curve for the piping material of interest.

In the original ASME pipe flaw evaluation procedures, this Z-factor was determined using the GE/EPRI J-estimation scheme for a circumferential through-wall flaw in the material of interest. The GE/EPRI estimation scheme was found to be the most conservative elastic-plastic analysis when compared to full-scale cracked-pipe tests [Brust et al., 1995]. The Z-factor is very sensitive to the material toughness and the pipe diameter and less sensitive to flaw size and the pipe's R/t ratio [Wilkowski et al., 1998]. In the ASME Code, the Z-factor is set equal to the maximum value of Z when the circumferential through-wall-crack length is varied. The Z-factor from these through-wall-cracked pipe analyses were then used in the ASME Section XI pipe flaw evaluation procedures for circumferential surface-cracked pipes [EPRI, 1986 and 1988].

Since that time, more detailed analyses have been conducted for circumferential surface-cracked pipe as well as comparisons to full-scale cracked-pipe tests at light water reactor (LWR) temperatures. The results of these analyses have shown that the ASME Z-factors for austenitic welds are reasonable approximations to the experimental results [Wilkowski et al., 1987]; however, the ferritic pipe and ferritic weld Z-factors results are very conservative [Wilkowski et al., 1996]. Current Code flaw evaluation procedures and Z-factors also do not exist for all flaw locations of interest (i.e., Inconel 82/182 weld metal and cast stainless steels more sensitive to thermal aging). Additionally, the effects of dynamic and cyclic loading for seismic conditions were not included in the ASME pipe flaw evaluation procedures, because (at the time the Z-factors were created) it was believed that the applied safety factors and use of Code strength properties compensated for such effects.

In the study reported herein, the staff calculated more realistic Z-factors for the following material cases:

   (1) A516 Grade 70 ferritic pipe base material
   (2) ferritic pipe submerged arc welds
   (3) Inconel 82/182 weld metal
   (4) Inconel 82/182 weld fusion line with ferritic steel
   (5) stainless steel submerged arc welds
   (6) cast stainless steel with different degrees of thermal aging

J-R Curves Used

The Z-factor corrects the limit-load solution to account for elastic-plastic fracture mechanics effect in the cracked-pipe problems for comparison to the limit load solution. The material fracture toughness is needed in the elastic-plastic fracture analysis, and the fracture toughness is characterized by the J-integral parameter, where the material resistance typically increases with crack growth by ductile tearing. The increasing toughness with crack growth is called the J-Resistance or J-R curve. The NRC's PIFRAC database, currently available at Emc^2 [Ghadiali and Wilkowski et al., 1996c], includes more than 800 J-R curves.

The typical J-R curve data provides the material toughness under quasi-static loading rates and monotonic loading, i.e., no dynamic or cyclic loading as would occur in a seismic event. Research efforts from the International Piping Integrity Research Group (IPIRG) programs (NRC and EPRI were members of that program), demonstrated the effects of dynamic loading and cyclic loading on the material toughness [Rudland, 1996]. The dynamic loading rate corresponded to a loading rate for a seismic event[‡‡] and recommendations for that rate have been used for material testing of Korean nuclear power plant piping steels [Wilkowski et al., 1998]. Generally, the austenitic piping materials have a slight increase in toughness at seismic loading rates, while about half of the ferritic base metals tested at seismic loading rates show a decrease in toughness (attributable to dynamic strain aging) [Rudland, 1996]. Ferritic steel weld data at seismic rates is sparse, but the few welds tested in the IPIRG program [Wilkowski et al., 1997] suggest that the toughness increases with seismic loading rates, probably as a result of a different dynamic strain-aging sensitivity than the base metals (carbon versus nitrogen dislocation pinning), but decreases with cyclic loading.

The cyclic effect on the J-R curve has also been evaluated in the IPIRG programs [Rudland, 1996] and [Marschall, 1989]. This cyclic loading during a seismic event is akin to interrupted ductile tearing. Detrimental effects on the toughness can occur when the reverse loading is compressive. The negative loading causes flattening of voids ahead of the crack tip and additional work hardening at the crack tip. The effects of cyclic loading are also sensitive to the material strength and the number and order of increasing large-amplitude cycles. Because the order of large amplitude cycles in a seismic event is very random, the decrease in the toughness from cyclic effects observed in the IPIRG program was reduced by half, i.e., the cyclic degradation factor was averaged with a value of 1 (no cyclic degradation).

[‡‡] The dynamic loading rate used for simulating seismic loading in J-R curve testing is to reach the maximum loading in one-quarter of the period of the first natural frequency, i.e., the frequency where the highest amplitude loading would be expected.

Table 4-8 gives the multipliers used on the monotonic-loaded quasi-static J-R curves of the various materials for dynamic and cyclic loading based on the trends from the IPIRG program and above comments. These multipliers are used because there are no applied safety factors for the best-estimate analyses, whereas the ASME Code SFs are believed to account for any potential detrimental effects. In some cases, the adjusted J-R curves are actually higher than the quasi-static J-R curve. The two ferritic materials had adjustments that made the seismic-adjusted J-R curve lower than the quasi-static J-R curve, which is attributed to dynamic-strain-aging effects commonly observed in these materials at LWR plant operating temperatures.

Table 4-8 Multipliers on Monotonic-Loaded Quasi-Static J-R Curves for Dynamic and Cyclic Loading Effects for Seismic Applications

Material	Dynamic effect multiplier	Cyclic effect multiplier[§§]	Total correction factor on quasi-static J-R curve for combined dynamic and cyclic loading
Inconel82 weld	1.2	0.9	1.08
Inconel82 fusion line	1.2	0.9	1.08
Stainless Steel SAW	1.2	0.9	1.08
Cast stainless steel with moderate sensitivity to thermal aging	1.2	0.9	1.08
Cast stainless steel with high sensitivity to thermal aging	1.2	0.9	1.08
A516 Grade 70 ferritic base metal	0.75	0.8	0.6
Ferritic submerged arc weld	1.5	0.6	0.9

The source of the quasi-static monotonic-loaded J-R curves is given below. All J-R curves were determined at 550°F to 600°F.

The Inconel 82/182 weld J-R curve is based on results presented in [Williams et al., 2004]. This was a weld removed from a cold leg for a plant that was not constructed in the 1980's [Wilkowski et al., 1989]. The J-R curve for Inconel 82/182 used in this analysis was an average of J-R curves that were very close to each other and were in the weld butter as well as main parts of the weld, and had crack growth in the radial or circumferential directions, i.e., the J-R curves did not seem to be very sensitive to the crack orientation or location in butter or main part of the weld. The modified J-R curve was used in the elastic-plastic fracture mechanics analysis (LBB.ENG2) rather than the standard deformation plasticity J-R curve because past analyses have shown better predictions of maximum loads in pipe tests when using the modified J-R curve [Wilkowski et al., 1998]. (The modified J-R curve is higher than the deformation plasticity J-R curve and, hence, is less conservative.)

[§§] Trend curves from the various nuclear piping materials tests with cyclic loading during J-R curve testing were used. The R-ratio (minimum/maximum loads) is a key parameter and was determined to be –0.6 using the linearly scaled 10^{-5} seismic stresses, i.e., (negative seismic stress amplitude plus normal stresses)/(positive seismic stress amplitude plus normal stresses). The calculated reduction Scale Factor from cyclic loading was averaged with a value of 1.0 due to uncertainties in the seismic load sequence and number of large amplitude cycles.

The Inconel 82/182 fusion-line J-R curve was also used because it gave a lower fracture resistance than the Inconel 82/182 weld metal and is a potential flaw location. This J-R curve came from large C(T) specimens developed during an investigation involving a full-scale pipe fracture test was conducted on a cold leg with a circumferential through-wall crack in the fusion line [Scott et al., 1995]. The modified J-R curve was used in this analysis because it gave the best predictions with the full-scale pipe test.

The stainless steel SAW J-R curve used in this analysis was based on a statistical analysis of data in the PIFRAC database [Ghadiali and Wilkowski, 1996c]. Those results showed no difference between the toughness of shielded metal arc welds and submerged arc welds, and were subsequently used as the technical basis for the current version of Section XI Appendix C that contains only one Z-factor equation for these two weld types. The mean minus one standard deviation J-R curve was used in these analyses because there was a statistically significant number of J-R curves. The modified J-R curve was used.

The ferritic A516 Gr 70 J-R curve was from Specimen F34-17 in the PIFRAC database [Ghadiali and Wilkowski, 1996c]. This curve was derived from the same cold-leg pipe from which the tensile tests data were taken, but was one of the lower J-R curves from that particular pipe. The modified J-R curve was used.

The ferritic pipe weld J-R curve was from Specimen F34-W31 in the PIFRAC database. This curve was derived from a weld made according to the specifications of a different NSSS supplier, and was one of the lower J-R curves from this weld. The modified J-R curve was used.

Cast stainless steels that are susceptible to thermal aging can have a much lower toughness than that for the stainless steel weld. This occurs only in a few plant cases. Analyses were not developed for severely aged cast stainless steel materials.

Z-Factor Analyses and Results

The Z-factors developed in this study were calculated using the LBB.ENG2 J-estimation scheme in Version 3.0 of the NRCPIPE computer code [Ghadiali, 1996a]. This code has been previously released to EPRI and members of the ASME Section XI Working Group on Pipe Flaw Evaluations. The LBB.ENG2 analysis is the most accurate circumferential through-wall-cracked-pipe J-estimation scheme when compared to experimental data [Brust et al., 1995]. The other data used in the analysis included the pipe base metal stress-strain curves (see the previous discussion), and the fracture toughness of the material as discussed above.

The Z-factors for the ferritic pipe base metals and welds are given in Figure 4-8 as well as the Z-factors from ASME Appendix C for reference. Note that despite the dynamic and cyclic toughness corrections on the J-R curves, the best-estimate Z-factors are significantly less severe than the current ASME Code Z-factors, especially for the 30- to 42-inch (76.2- to 106.7-cm) diameter pipes being evaluated.

The Z-factors for the austenitic pipe base metals and welds are given in Figure 4-9 as well as the Z-factors from ASME Appendix C for reference. The stainless steel weld Best-Estimate Z-factor is lower than the ASME Code Z-factor for the 30- to 36-inch (76.2- to 91.4-cm) diameter pipes of interest. The Inconel weld metal and fusion-line Z-factors are less severe than those for the stainless steel weld.

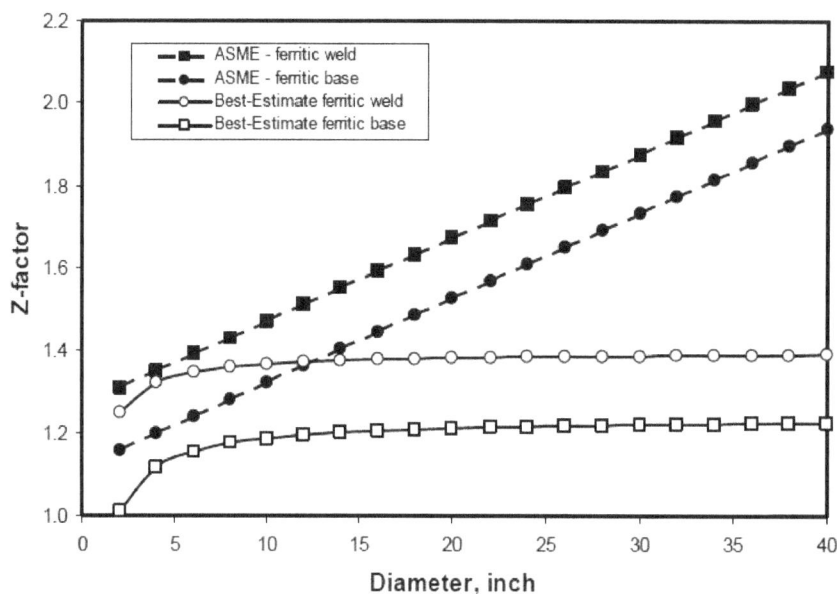

Figure 4-8 Comparison of ASME Code and Best-Estimate Z-Factors for Ferritic Steels in Plant C Cases

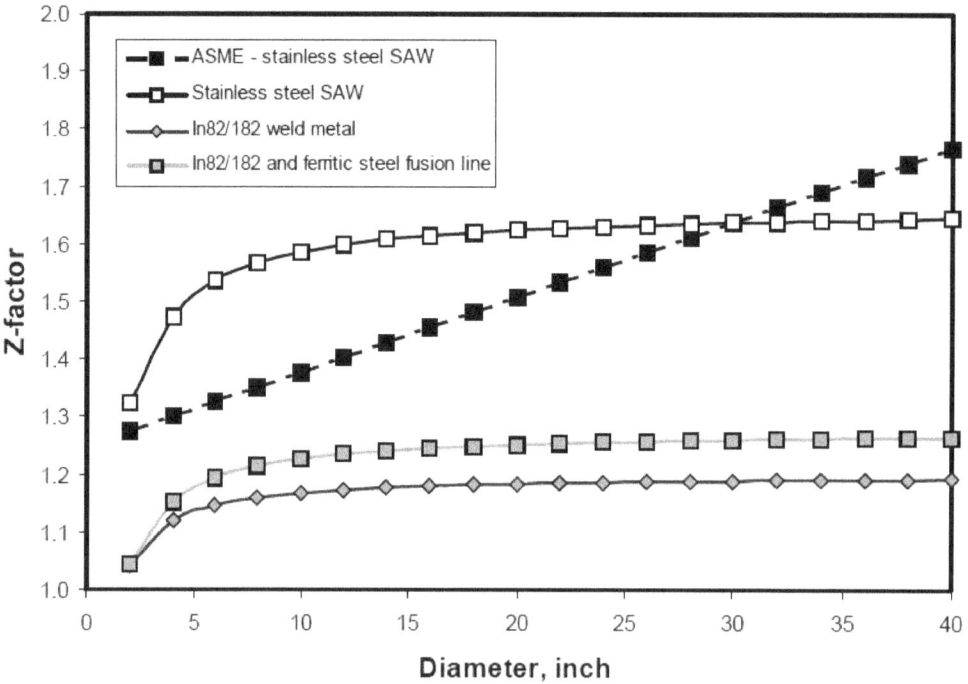

Figure 4-9 Comparison of ASME Code and Best-Estimate Z-Factors for Austenitic Base Metals and Welds in Plant A and B Cases and the Plant C Cold-Leg Case
(Best-estimate Z-factors based on through-wall-crack elastic-plastic fracture mechanics analyses)

4.5.2.8 Calculate the Critical Surface Flaw Size at 10^{-5} or 10^{-6} Annual Probability of Exceedance Seismic Event

In this final step, the critical flaw size is calculated for the N+10^{-5} stresses using values from the previous five steps and with the safety factor equal to one. This flaw size is plotted in terms of the circumferential surface flaw depth normalized by the thickness (a/t) versus the circumferential length of the flaw normalized by the pipe circumference (θ/π). This curve is then compared with the ASME Code allowable flaw sizes at the design stresses developed in accordance with Section 4.5.2. This entire process is illustrated by the example in the following section.

4.5.2.9 Application Example for Surface Flaw Evaluation Procedures

To demonstrate the application of the flawed pipe evaluation procedure in Figure 4-6, the staff selected an example of hot-leg piping from a four-loop Westinghouse plant (Plant "A"). This is austenitic steel piping with an inside diameter of 29.2 inches (74.2 cm) and a thickness of 2.37 inches (6.0 cm). The plant is located on a rock site. This example is only for a 10^{-5} annual probability of exceedance seismic event case, but the analysis for 10^{-6} annual probability of exceedance seismic event would be similar.

From the LBB database, the normal operating stress and SSE stresses for this piping are about 10 and 15 ksi, respectively, at the nozzle that is at the highest stress location. From the LLNL seismic hazard curve for this plant, the ratio of PGA at 10^{-5} annual probability of exceedance to the SSE is about 3.2. This value is used as a multiplier on the SSE component of the stress to obtain the elastic stresses at the 10^{-5} annual probability of exceedance seismic event. This elastic stress, without any correction factors, is about 50 ksi.

The next step is to reduce the linearly scaled seismic stresses by the Scale Factor discussed in Section 4.4.2. This correction factor accounts for; (1) shape differences between uniform hazard spectra (UHS) and design spectra used, (2) differences between the damping of the structures from design and best-estimate values, and (3) differences in soil-structure interaction analysis methods between design and best-estimate values, if applicable. For this plant, the Scale Factor was 0.49. Hence, the corrected stress for the 10^{-5} annual probability of exceedance seismic event is about 24 ksi.

A nonlinear stress correction is then applied to the total stress. The total stress is the normal stress plus the corrected stress for the 10^{-5} annual probability of exceedance seismic event, or about 34 ksi. In this evaluation, the nonlinear correction was determined from the relationship illustrated in Figure 4-7. The correction factor is 0.885, so that the total stress is about 30 ksi (Section 4.5.2.5). This includes a pressure-induced longitudinal membrane stress of 7.44 ksi for the operating pressure of 2,250 psig.

With these stress values, a separate spreadsheet was created to calculate the allowable flaw length associated with a particular flaw depth. The safety factors were set equal to 1.0 for the critical flaw size analysis, and the Z-factors from Figure 4-9 were used. The Z-factor is determined from the material and mean pipe diameter. The pipe inside diameter is 29.2 inches (74.2 cm) and the thickness is 2.37 inches (6.0 cm), giving a mean diameter of 31.57 inches (80.2 cm). The best-estimate Z-factor for a stainless steel SAW (Figure 4-9) in this pipe is 1.65. The more realistic actual strength values give a flow stress of 52.95 ksi which is 22.6% higher than using the Code values from Section II for this material. The more realistic strength values were used in the best-estimate analysis for critical flaw size evaluation for the 10^{-5} annual probability of exceedance seismic event. For the Code evaluations, the yield and ultimate strengths for CPF-8A pipe were used at 600°F (pages 535 and 443 in Section II – Part D).

Figure 4-10 shows results of the analysis and compares best-estimate versus Code-allowable flaws. For this piping system case, the most typical limiting material condition is the ASME Code stainless steel SAW maximum allowable flaw sizes and the best-estimate stainless steel SAW critical flaw sizes. For that situation, the ASME curve is significantly below the best-estimate curve, so that the Section XI flaw evaluation procedures should prevent flaws from developing to the critical size for the 10^{-5} annual probability of exceedance seismic event.

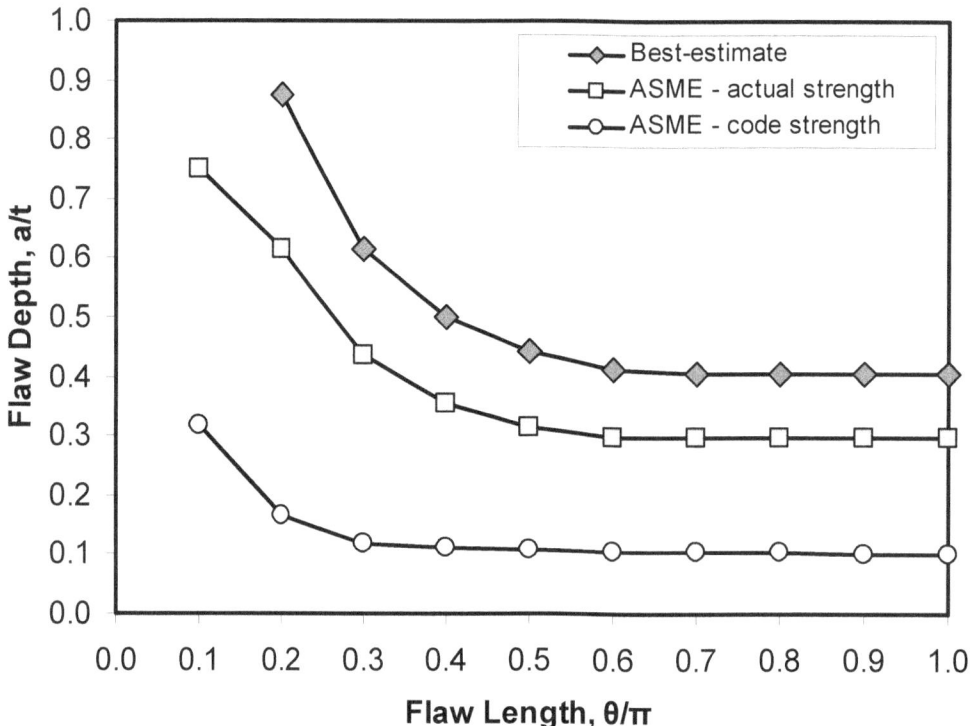

Figure 4-10 Comparison of ASME Maximum Allowable Flaw Sizes to Best-Estimate Critical Flaw Sizes at 10^{-5} Annual Probability of Exceedance Seismic Event for Plant A Hot-Leg Case

4.5.2.10 Surface-Flawed Piping Analysis Results

The above procedure was applied primarily to rock-site PWR plants east of the Rocky Mountains for which SSE stresses were available from the LBB database. As previously stated, only large-diameter piping in PWRs was considered, i.e., hot leg, cold leg, and cross-over legs. In total 52 piping system cases from 27 plants were analyzed to cover a range of stresses and material conditions.

The analysis considered only flaws in stainless steel welds for the Westinghouse plants with stainless steel primary piping, and only flaws in the ferritic welds for the primary piping in the CE plants.

Note that the nonlinear correction factors were small in the best-estimate analysis for the 10^{-5} annual probability of exceedance seismic events because the stresses are not that much above the yield strength of the material, but would be much more significant for the 10^{-6} annual probability of exceedance seismic events. Although, best-estimate analyses were not conducted for flaws in Inconel welds and fusion lines (which the Code does not currently have), this material location is very tough and flaws in the stainless steel weld metal would be controlling, except where aging-sensitive cast stainless steels exist.

The staff conducted the following four specific analysis procedures for each of the 52 piping systems:

(1) ASME allowable flaw size analysis based on actual strength properties

(2) ASME allowable flaw size analysis based on Code strength properties

(3) critical flaw size analysis for a 10^{-5} annual probability of exceedance seismic event based on actual strength properties

(4) critical flaw size analysis for a 10^{-6} annual probability of exceedance seismic event based on actual strength properties

Each of these analyses produces a locus of flaw depths and flaw lengths that are regarded as being allowable for the ASME analysis, but are regarded as being critical for the 10^{-5} and 10^{-6} annual probability of exceedance seismic event analyses. Figure 4-10, discussed above, shows results typical of any of these analyses and illustrates how the results of Procedures 1, 2, and 3 may appear relative to each other. Procedure 4 was conducted to investigate sensitivity to higher stresses and available margins. The 52 pipe system cases analyzed exhibited the following three categories of outcomes:

Category A. The critical flaw sizes estimated for 10^{-5} or 10^{-6} annual probability of exceedance seismic event (Procedure 3) are larger than the flaw sizes that are currently allowed by Section XI of the ASME BPV Code, using both Code and estimated actual material properties (Procedures 1 and 2, respectively). Figure 4-11 illustrates this situation, which is the most common situation for the 10^{-5} annual probability of exceedance seismic event.

Category B. The critical flaw sizes estimated for 10^{-5} or 10^{-6} annual probability of exceedance seismic event lie between Procedures 1 and 2. That is, the critical flaw sizes based on Procedure 2 are larger than those based on Procedure 3, while those based on Procedure 1 are smaller. Figure 4-12 illustrates this situation.

Category C. The critical flaw sizes estimated for 10^{-5} or 10^{-6} annual probability of exceedance seismic event are smaller than the flaw sizes for both Procedures 1 and 2, as illustrated in Figure 4-13.

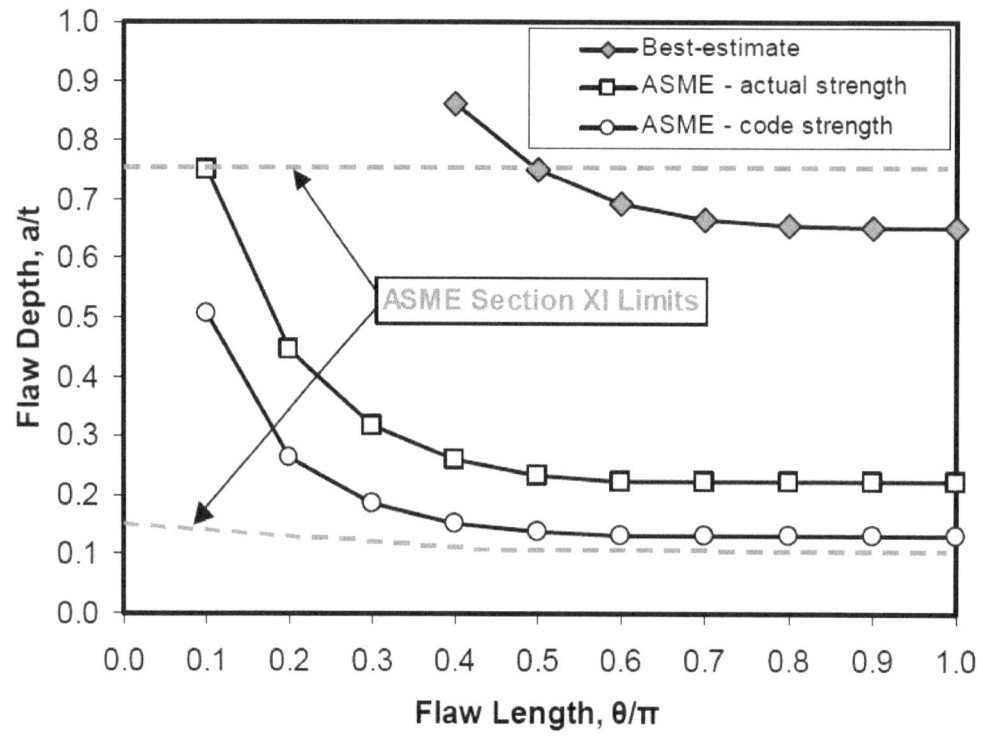

Figure 4-11 Example of Category A Results for a Hot Leg

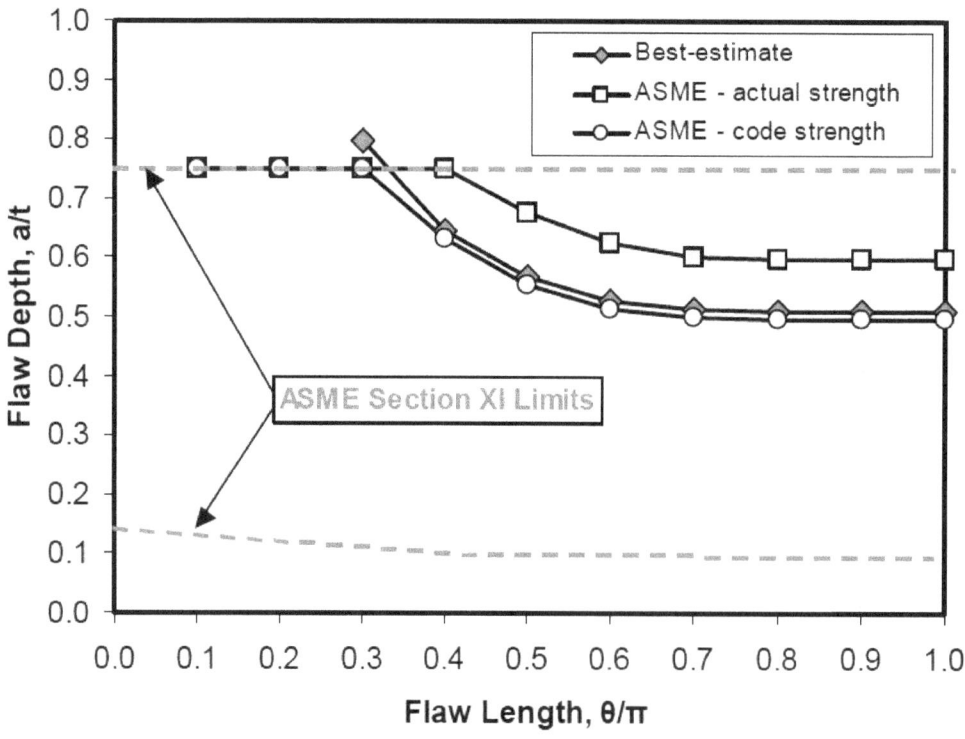

Figure 4-12 Example of Category B Results for a Crossover Leg at the 10^{-5} Annual Probability of Exceedance Seismic Event

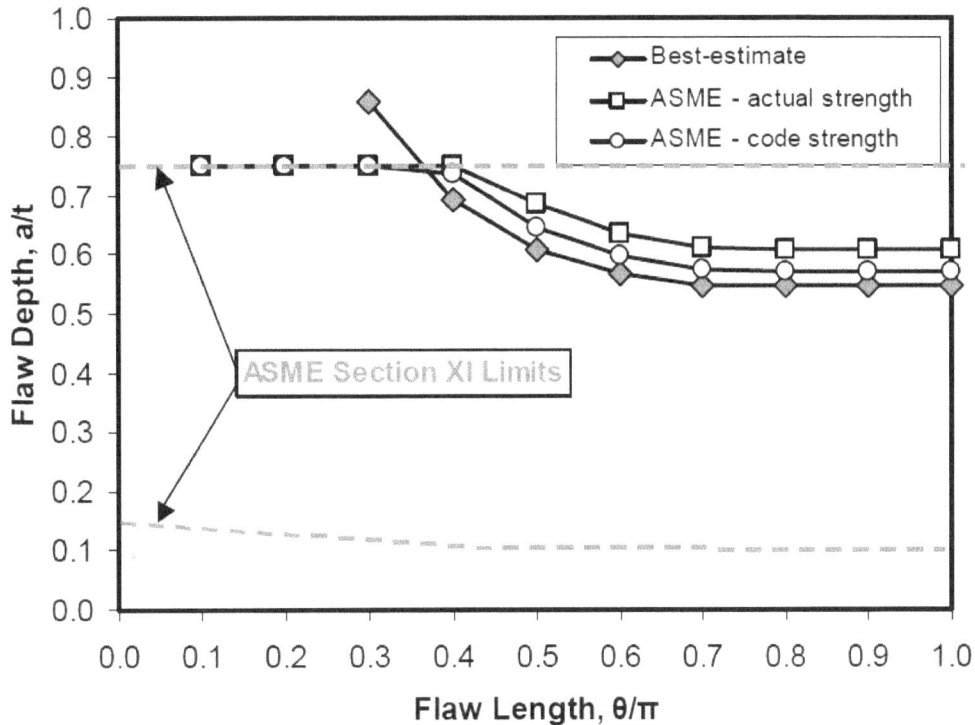

Figure 4-13 Example of Category C Results for Cold Leg on the Discharge Side of a CE Plant at the 10^{-5} Annual Probability of Exceedance Seismic Event

Several observations are made with respect to the above figures and results of the 52 analyses:

- For the 10^{-5} annual probability of exceedance seismic event, the results of 48 analyses belong to Category A, 1 falls into Category B, and 3 are in Category C. This breakdown changes significantly for the 10^{-6} annual probability of exceedance seismic event, with the results of 20 analyses in Category A, 20 in Category B, and 12 in Category C.

- Results in Category A suggest that the current ASME Section XI flaw evaluation and ISI requirements would preclude the possibility of flaws growing to a critical size and leading to a seismic-induced break.

- The use of actual material properties has an appreciable effect on allowable flaw sizes. The results are biased by whether a licensee uses nominal or actual material strengths in its flaw evaluation procedure.

Regardless of whether results belong to Category A, B, or C, the following observations provide broader perspective on the seismic behavior of flawed piping.

As seen in Figure 4-11 through Figure 4-13, even for very long circumferential flaws, the minimum critical a/t ratio is about 0.5 for a 10^{-5} annual probability of exceedance seismic event. For shorter length circumferential flaw the critical a/t ratios become quite large. These facts are further highlighted in Figure 4-14 and Figure 4-15. Figure 4-14 shows the critical a/t value for N+10^{-5} stresses for very long circumferential flaws represented by θ/π value of 0.80. The minimum a/t values for this large flaw is about 0.4 and it is substantially greater for many cases.

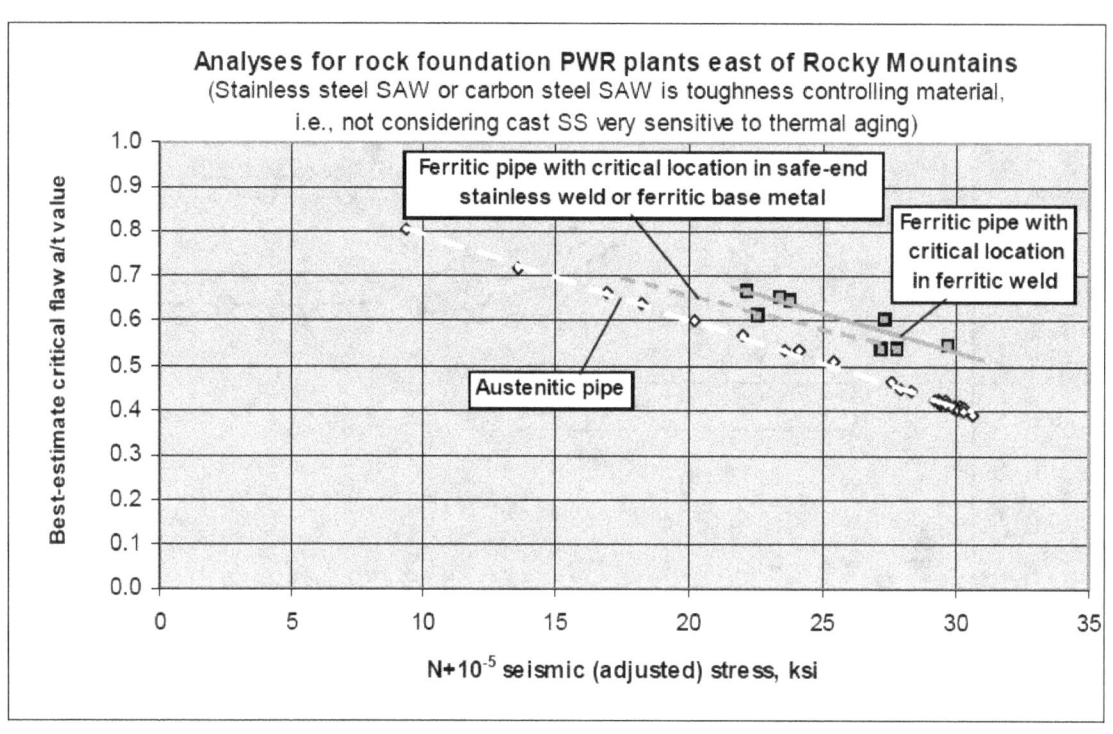

Figure 4-14 Critical Surface Flaw a/t Values Versus N+10^{-5} Stresses for θ/π = 0.8 (For all loops and plants and all materials together)

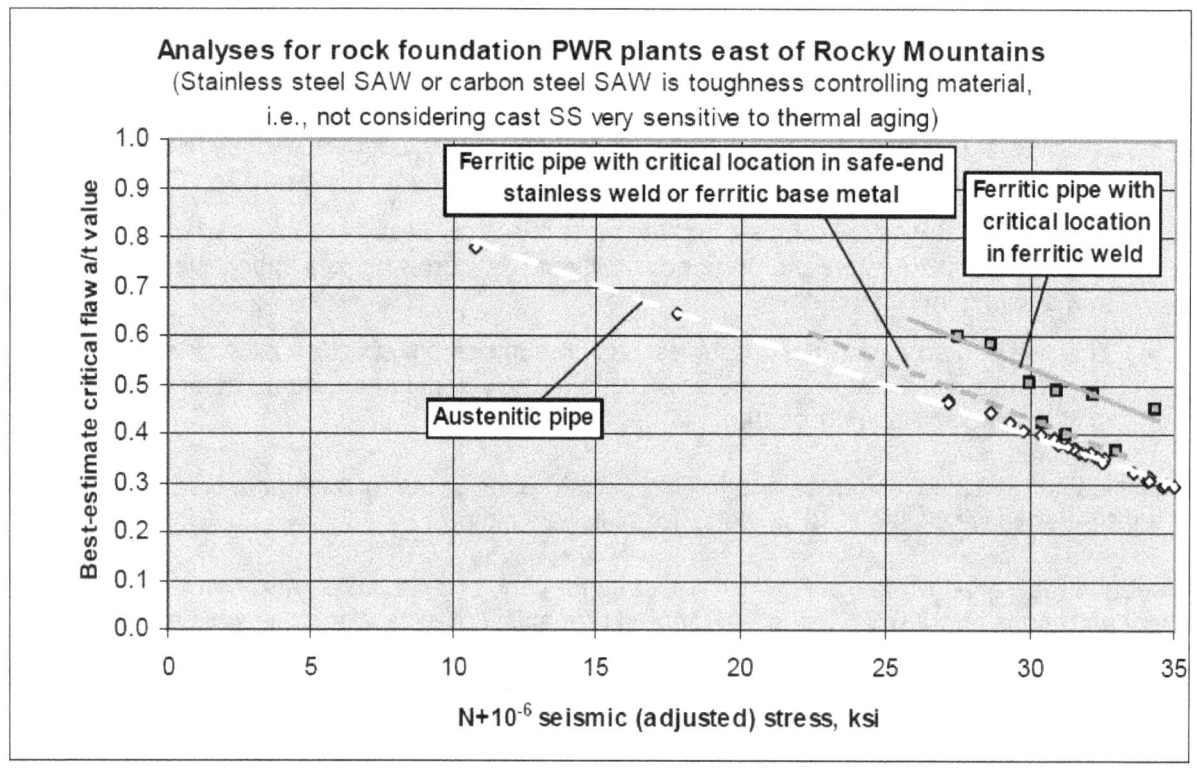

Figure 4-15 Critical Surface Flaw a/t Values Versus N+10^{-6} Per Year Seismic Stresses for θ/π = 0.8 (For all loops and plants and all materials together)

Figure 4-15 shows critical a/t value for N+10^{-6} stresses for the same long circumferential flaw. Again, the minimum a/t ratio is 0.3. By contrast, the Code-allowable a/t ratio approaches values as small as 0.1 in several cases.

Figure 4-16 shows circumferential crack lengths associated with the critical through-wall cracks. At high stresses these lengths are on the order of about 15% of the pipe circumference.

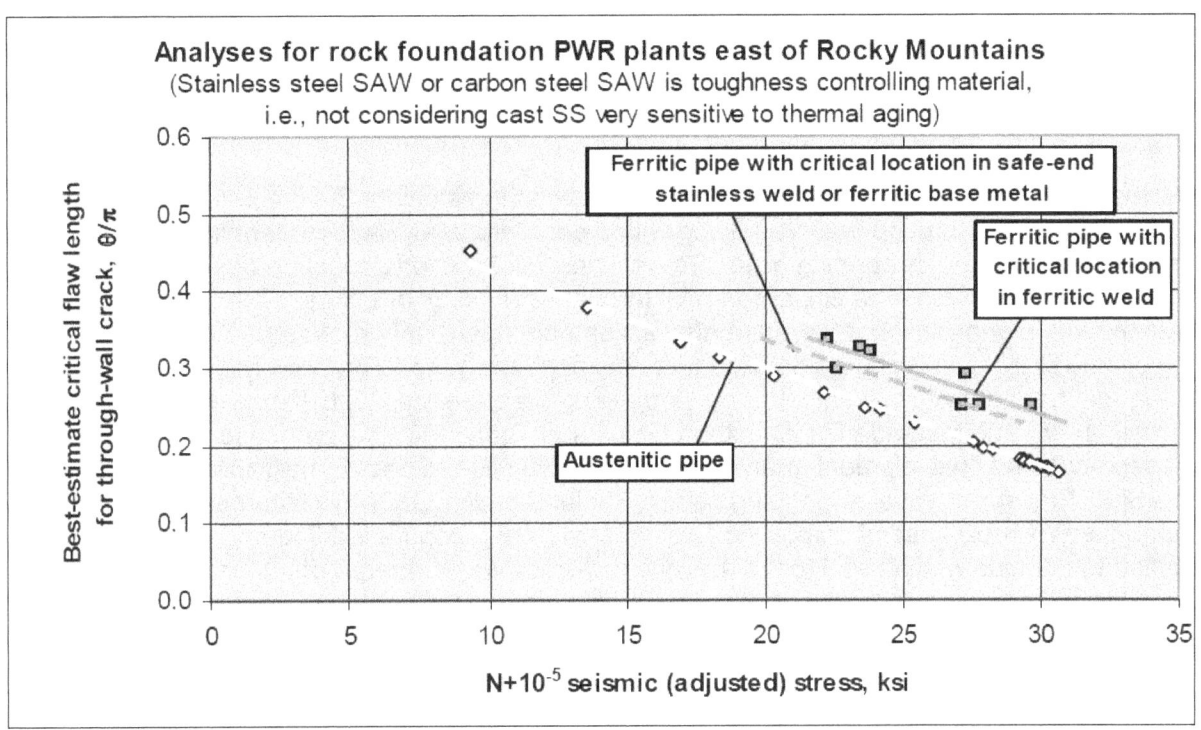

Figure 4-16 Critical Circumferential Through-Wall Flaw Length Corresponding to 10^{-5} Per Year Seismic Stresses

4.5.3 Through-Wall Flaw Evaluation Procedures for LBB Analysis

The second approach is similar to an LBB analysis, but for critical circumferential through-wall cracks at the 10^{-5} and 10^{-6} seismic stresses. The general analysis procedures involve first calculating the through-wall flaw length that can be detected by leakage with appropriate safety factors, and comparing that flaw length to the critical through-wall flaw length. Note that many of these items are discussed in greater detail [Scott, 2002], which provides a technical basis for a potential regulatory guide on LBB, which might replace the draft SRP 3.6.3.

4.5.3.1 Determination of the Leakage Flaw Size

The leakage calculation requires determining the crack-opening displacement (COD) for the applied normal operating loads in a pipe with a circumferential through-wall crack, and then determining the leakage through that crack for the given crack morphology parameters (roughness, number of turns, actual flow path relative to the pipe thickness, etc.). To determine the crack size for a given leak rate involves an iterative analysis, and in this report the analyses used a leak-rate computer code called "Seepage Quantification in Reactor Tubing" (SQUIRT). This code was recently updated into a Microsoft Windows® version with many new deterministic improvements [Ghadiali, 2003], and provided to industry for their recent LBB evaluations for PWSCC. All analyses in this project were done using this latest update to the SQUIRT code.

There are a number of uncertainties associated with leakage detection. These consist of the uncertainties in the leakage monitoring equipment, uncertainties in calculating the crack opening area, and uncertainities in the thermal-hydraulic analyses. Two NUREG-series reports, by Rahman [1995a] and Ghadiali et al. [1996b], provide analyses associated with defining some of the uncertainties in LBB analyses. Some of these uncertainities are described later and show the need to have a safety factor on leak rate.

Crack-Opening-Area Analyses

The leakage flaw size is determined by using a method to calculate the crack-opening area for the circumferential through-wall flaw along with a separate analysis for determining the leakage through that given crack opening area. Several analysis methods could be used for determining the crack-opening area for a circumferential through-wall crack in a pipe. One method is finite element stress analyses, but there are also acceptable analytical procedures that have been benchmarked against the more detailed finite element analyses. The two analytical methods used most frequently to calculate the crack-opening area are the GE-EPRI estimation procedure [Kumar, 1986] and the Tada-Paris method [Paris and Tada, 1983]. Comparisons were made between detailed finite element analyses and the GE/EPRI method or the Tada-Paris method in an ASME Pressure Vessel and Piping paper [Rudland et al., 2000]. This comparison was for the relatively simple case of combined axial tension from pressure loading and global bending applied to the pipe with no end restraints, and the crack was centered on the bending plane with no welds. Those results showed that the GE/EPRI method was much more accurate than the Tada-Paris method and in close agreement with the finite element results as long as the applied stresses were less than 60-percent of the failure stress. This 60-percent limit is readily satisfied in LBB analyses. Hence, in the following leakage analyses in this report, only the GE/EPRI COD analysis was used. Additionally, it was found in the finite element analyses that the shape of the crack opening was elliptical, so that by knowing the COD at the center of the flaw and the length of the flaw, the crack opening area could be calculated for the leakage.

Some additional contributions to the uncertainities in calculating the crack opening area are:

- It is always assumed that the crack is centered on the bending plane of the normal operating stresses. The leakage size crack would be longer if it was not perfectly centered on the bending plane. Some analytical results have been developed for the off-centered crack, but are not applied in this report [Rahman, 1995b].

- The effects of weld residual stresses are typically ignored in the analyses. The weld residual stresses can cause the crack faces to rotate and affect the crack opening on the ID surface (see Scott et al., 2005, Appendix H). These are not large effects for thick-walled pipe (without repair welds) and were ignored in the analysis in this report.

- Thickness differences at a nozzle location on one side of the crack can cause a smaller COD [Rahman, 1995b]. Again, these effects were not included.

- The pipe-system boundary conditions are different than the simple free-ended pipe axial tension and bending analyses in the GE/EPRI or other analyses methods. As the crack gets long then the end conditions, i.e., the distance from the crack to a fixed end like a nozzle, can restrain any induced bending from the axial membrane loads (see [Scott et al., 2005, Appendix D]. This effect is again reasonably small for large-diameter pipes, where the leakage crack lengths are a small percent of the entire circumference and, hence, was ignored in this analysis.

The combined uncertainty from the above four technical aspects are covered by the safety factor on leakage.

Thermohydraulic Analyses

The actual leakage calculations are done using a leak-rate code that accounts for two-phase flow through the crack. The basic analysis is typically the Henry-Fauske analysis for flow through tubes. This analysis is then modified for an elliptical cross-section tube in the SQUIRT Code [Paul, 1994] and EPRI's PICEP code [Norris et al., 1984]. Such codes are coupled to subroutines with the steam tables for water. The SQUIRT code has been compared to numerous idealized leak-rate tests [Scott et al., 2003]. Generally the accuracy is plus or minus a factor of 2 in the leak rate ranges of interest in this report.

One of the critical aspects in the leak-rate analysis is the type of crack mechanism, which determines the surface roughness, number of turns, and actual flow path length relative to the pipe thickness. The work by Rahman [1995a] and Wilkowski et al. [2005] statistically defined these parameters for IGSCC cracks, corrosion-fatigue cracks, air fatigue cracks, and PWSCC cracks found in service. In this report, the mean values for those parameters were used. Rahman [1995a] showed that going from the mean value to the 2-percent fractile level of these crack morphology parameters accounted for a factor of 2 on leak rate. Hence, the safety factor on leak rate needs to be great enough to account for these uncertainties, as well as the uncertainties in the crack opening area evaluation and those from the accuracy of the measurement systems. A factor of 10 has been used to account for all of these uncertainities.

Finally, another aspect that is an interaction between the crack-opening structural calculation and the leak rate is that the effective roughness and number of turns depends on the COD. An example of this effect is easily envisioned for an IGSCC crack, where for a crack with a very large COD across the crack faces, the roughness is approximately equal to one-half of the diameter of the grains in the stainless steels. For the same crack when it is very tight, then the roughness is equal to that of the grain boundaries, and there are a large number of turns. Additionally, the tight crack has a much longer flow path than just the thickness of the pipe. The statistical analysis of the crack morphology parameters from the different cracking mechanisms was determined for tight cracks as well as cracks with large crack openings by examining metallographic sections of cracks removed from service. This COD-dependant crack morphology approach was first developed by Rahman [1995a] and incorporated in the SQUIRT code as an option, with some additional improvements in the report by Y. Chen [in Kupperman, 2004]. The consideration of the effects of the COD on the effective roughness and number of turns is very important in probabilistic analyses because the leak rate codes will have numerical instabilities if the COD is less than or equal to the surface roughness. A few spot checks were made on this effect, and the COD was generally large enough that it did not matter if the COD-dependant analysis was used.

Critical Through-Wall-Crack Lengths

The critical circumferential through-wall crack lengths were calculated for the 52 different PWR pipe systems previously analyzed in the surface-crack evaluations. This was done for the 10^{-5} and 10^{-6} normal plus SSE stresses with the various corrections to the seismic design stresses and the nonlinear correction to the total stress as previously described. Figure 4-16 presented the critical through-wall flaw results for N+10^{-5} seismic loading using the best-estimate Z-factor approach. The result for the N+10^{-6} seismic loading is given in Figure 4-17. As with the prior surface flaw evaluations, these analyses considered the stainless steel weld and ferritic welds to be the bounding toughness location for the cracks. Cast stainless steels sensitive to thermal aging could have a lower toughness, and are not included in this evaluation.

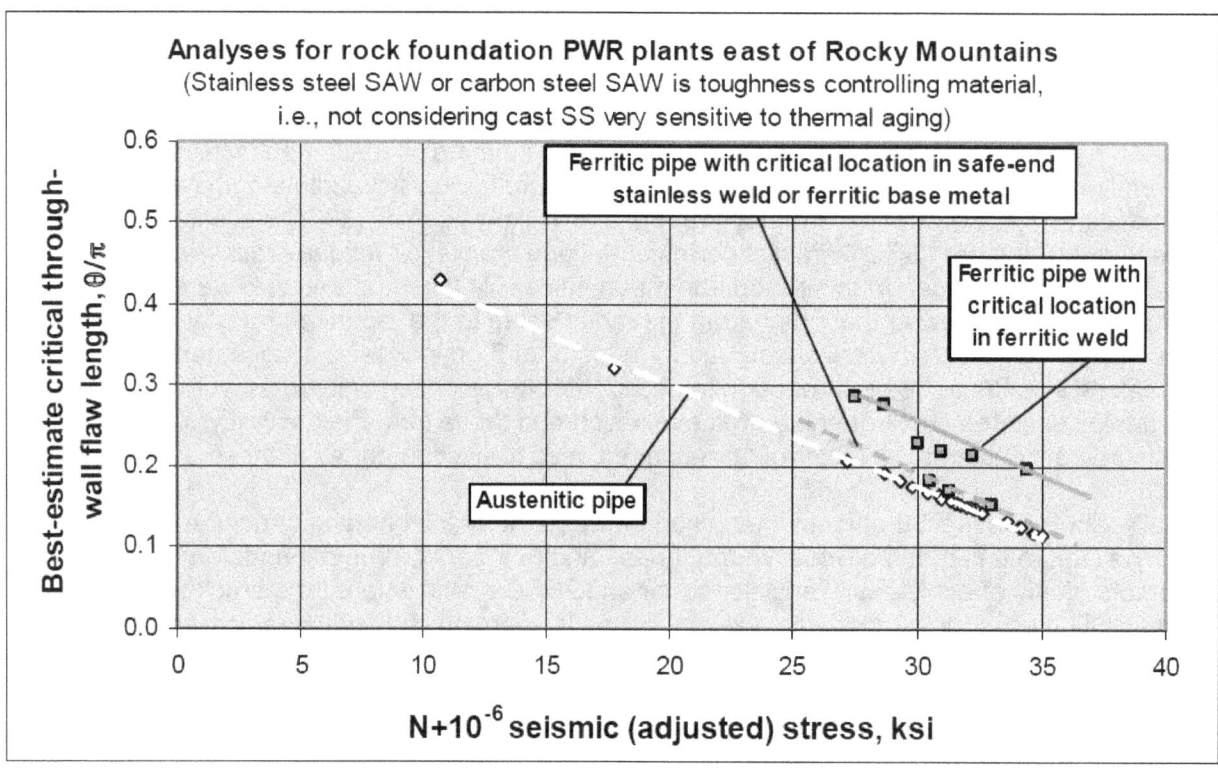

Figure 4-17 Critical Circumferential Through-Wall Flaw Length Corresponding to 10^{-6} Per Year Seismic Stresses

4.5.3.2   Examples of LBB Analyses and Results

Of the 52 different PWR pipe systems analyzed, Figure 4-16 and Figure 4-17 showed that the stainless steel weld cases had the lowest critical through-wall flaw sizes. Consequently, several of those pipe systems were selected for these LBB analyses.

In these analyses three different cracking mechanisms were considered:

(1) air fatigue crack
(2) corrosion-fatigue crack
(3) PWSCC crack

The air fatigue crack is very smooth with very few or negligible turns through the thickness, which gives the shortest through-wall-crack length for a given leak rate. This was used in many of the past LBB submittals.

The corrosion-fatigue crack is rougher than the air-fatigue crack and gives a longer through-wall crack length for a given leak rate. The corrosion-fatigue crack has been suggested for use in the LBB submittals [Scott, 2002].

The PWSCC crack was used because such cracks have recently occurred in PWR plants in Inconel 82/182 bimetallic welds. Such welds exist in the large-diameter pipe systems under consideration for the 10 CFR 50.46 rule change. A limited number of PWSCC cracks removed from service were evaluated to determine the PWSCC crack morphology parameters [Wilkowski et al., 2005]. A PWSCC crack that grows parallel to the dendritic grains will have more severe crack morphology parameters than a corrosion-fatigue crack, but less severe than an IGSCC crack in BWR piping. This crack growth direction occurs if the PWSCC is growing radially in the main part of a bimetallic weld. A PWSCC crack that is growing perpendicular to the dendritic grains will have a much more severe crack morphology, which corresponds to a crack growing radially, but in the buttered region of the bimetallic weld.

Figure 4-18 shows the results of a sensitivity study for several plant piping systems where the technical specification leakage of 1 gpm was used and the applied safety factor on the leak rate was 10. There was **no** safety factor on the leakage size crack in this evaluation, in contrast to the safety factor of 2 used in the draft SRP 3.6.3 for LBB analyses for stresses equal to N+SSE. This figure shows the leakage size crack relative to the critical through-wall-crack length at the 10^{-5} seismic loading versus the normal stresses divided by the normal plus 10^{-5} seismic stresses. Several interesting aspects can be seen from this evaluation:

(1) The leakage size flaws are less than the critical size flaws when no safety factor is used on the crack length.
(2) As the N/(N+10^{-5} seismic) stress ratio decreases (at lower normal operating stresses or higher seismic stresses), the leakage flaw size is closer to the critical flaw size.
(3) The PWSCC crack lengths are about 50 percent longer than the air fatigue cracks for the same leakage.
(4) The corrosion-fatigue crack case is closer to the PWSCC crack case than the air fatigue crack.

It should be noted that of the 43 stainless steel pipe systems examined, the range of the N/(N+10^{-5} seismic) stress ratio was from 0.25 to 0.91 with an average value of 0.59.

As a sensitivity study, a Safety Factor of 1.5 was evaluated for the 10^{-5} seismic loading. These results are shown in Figure 4-19, which shows two curves. One is for the traditional technical specification leakage of 1 gpm. The second curve shows what happens if the 0.5-gpm leak rate was used. That leak rate was acceptable in a few recent LBB submittals. Hence, with this leak rate and the 1-gpm technical specification leakage, most of the analyzed cases satisfy the Safety Factor of 1.5.

A final analysis was done for the 10^{-6} seismic loading with all the same correction factors applied to the seismic stresses as presented earlier. The difference here is that there is a safety factor of 1.0 on the leakage flaw size when comparing it to the critical flaw size. These results are shown in Figure 4-20. As with the 10^{-5} seismic loading case, there are a few cases that will not meet this criterion with the technical specification leakage of 1 gpm. However, with the leakage detection capability of 0.5 gpm, the safety factor is greater than 1 for the 10^{-6} loading.

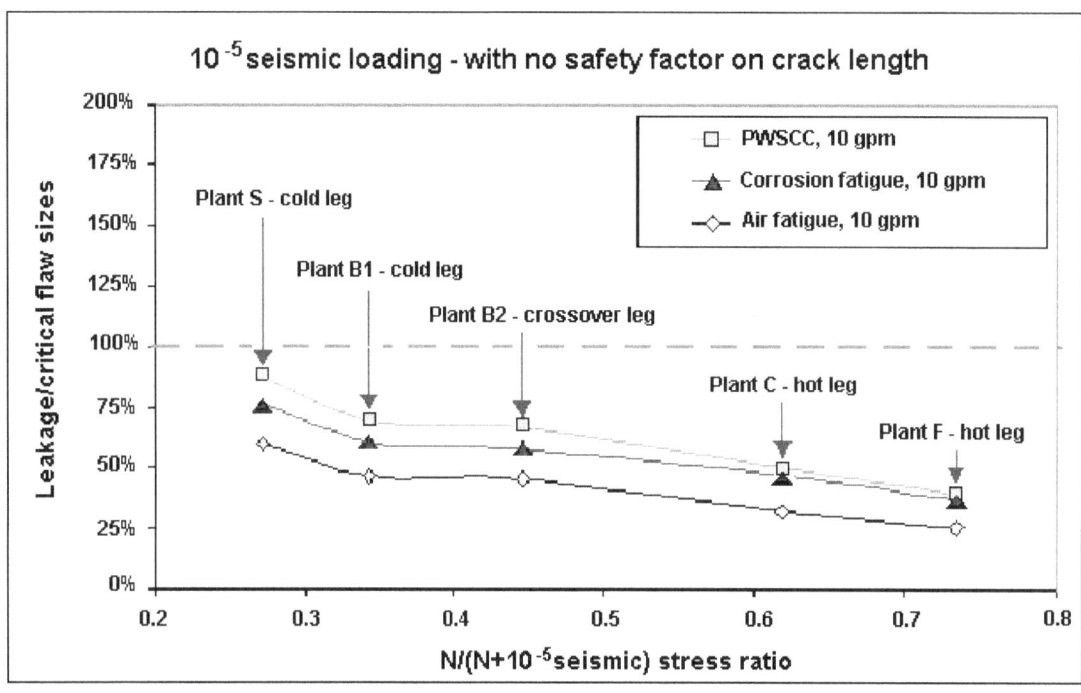

Figure 4-18 Sensitivity Study Results Showing Leakage to Critical Crack Size Ratio Versus Normal to N + 10^{-5} Seismic Stress Ratio for Five Different Plant Piping Systems with Different Cracking Mechanisms

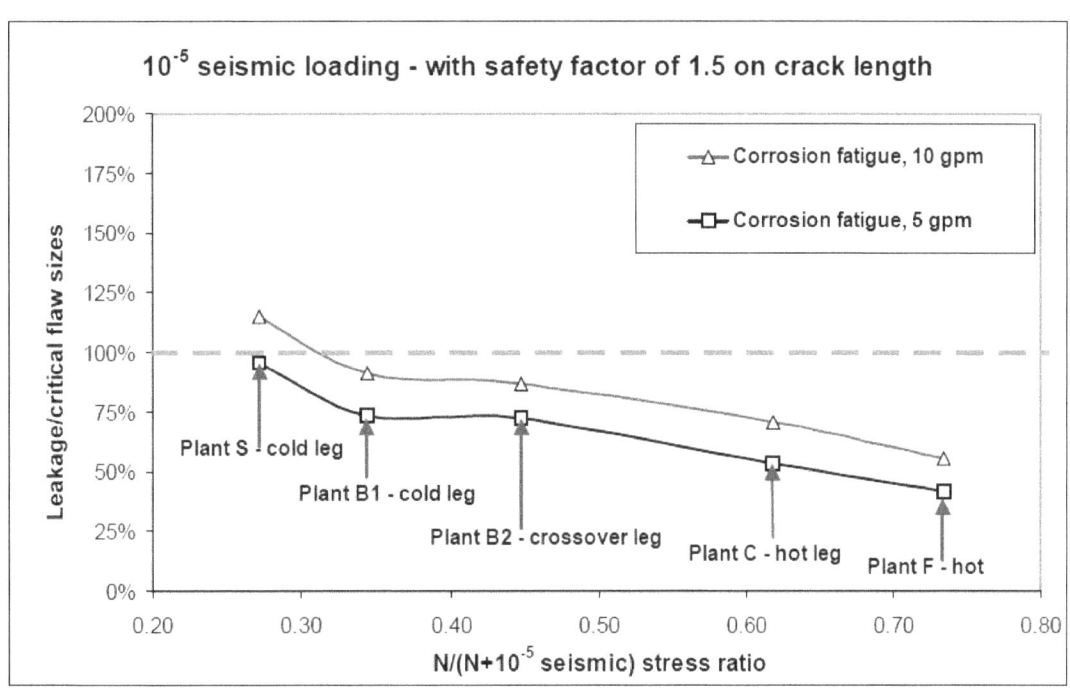

Figure 4-19 Sensitivity Study Results Showing Leakage to Critical Crack Size Ratio Versus Normal to N + 10^{-5} Seismic Stress Ratio for Five Different Plant Piping Systems with Safety Factor of 1.5 on Crack Length and Safety Factor of 10 on Leak Rate

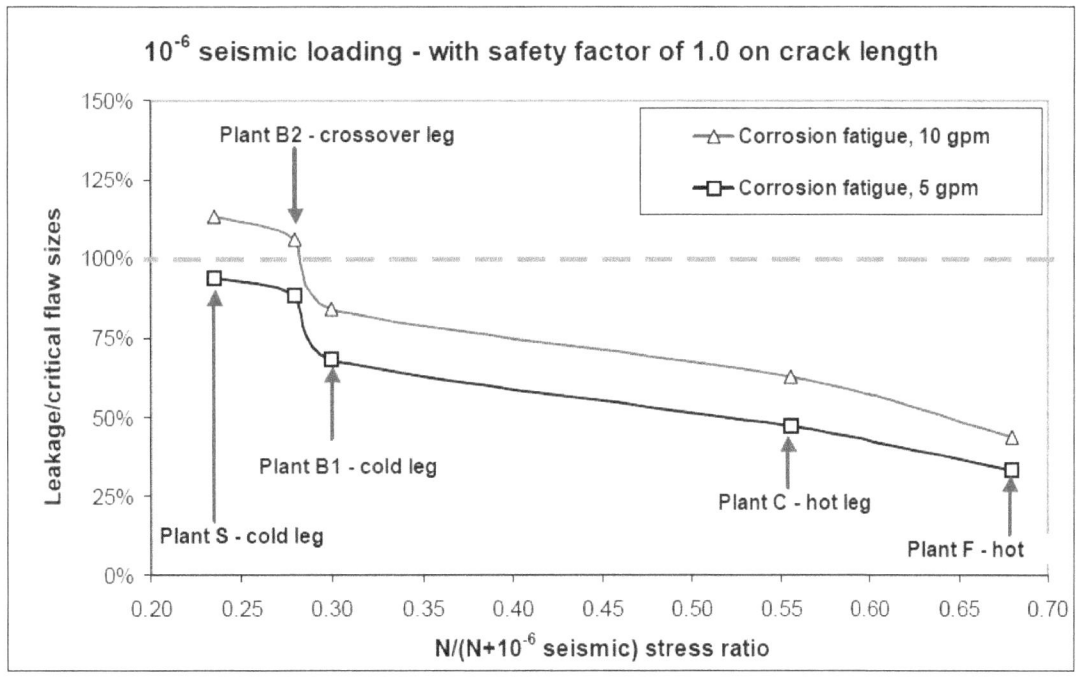

Figure 4-20 Sensitivity Study Results Showing Leakage to Critical Crack Size Ratio Versus Normal to N + 10^{-6} Seismic Stress Ratio for Five Different Plant Piping Systems with Safety Factor of 1.0 on Crack Length and Either the 1 gpm Technical Specification Leak Rate or 0.5 gpm Leak Rate (both with a safety factor of 10)

4.5.4 Overall Results of Flawed-Piping Analysis

The overall implication of the results and observations in Sections 4.5.2.10 and 4.5.3.2 is that very large flaws need to be present to produce seismic-induced breaks even for very rare / large earthquakes. The probability of having such flaws in service is mitigated by considering the following factors:

(1) Inservice inspections are generally capable of detecting much smaller flaws.

(2) The past practices have generally been intended to repair a flaw on discovery for most of the materials used in primary loop piping, or to mitigate conditions that lead to flaw growth.

(3) Analyses conducted using an approach similar to that used for LBB evaluations have generally demonstrated that leakage flaw sizes are much smaller than critical flaw sizes, and detectable leaks can occur before flaws reach critical size. At high stresses, these lengths are on the order of about 15% of the pipe circumference, while in LBB analyses, typical leakage flaw sizes in PWR primary piping are about 6% of the circumference. Hence, for degradation mechanisms where long surface cracks do not develop, some protection would naturally occur from leakage detection. However, leakage protection may not necessarily occur for flaws developed from stress-corrosion cracking mechanisms where long surface cracks could develop.

4.6 Indirectly Induced Piping Failures

4.6.1 Overview and Background

As discussed in Section 3.4, the staff's approach for considering indirect failures is based on the LLNL study [NRC, 1985a, 1985b, 1985c, 1988a]. Indirect failures are pipe ruptures caused by failures of major components or component supports as a result of an earthquake.

In the present study, the staff conducted a scoping evaluation of indirect failure of the RCL piping systems attributable to structural failure of RCL major components or their supports. This scoping evaluation used assumptions and data available in various NUREG-series reports [NRC, 1985a, 1985b, 1985c, 1988a]. Similar to other parts of this study, this evaluation focused on PWRs, and considered Westinghouse and CE as the NSSS vendors. For indirectly induced failure, detailed design information was not available for the RCL or its major components (reactor pressure vessel, steam generators, and reactor coolant pumps). The seismic hazard was derived from the seismic hazard curves and UHS in NUREG-1488 [Sobel, 1994], and the staff developed structure response factors based on available information. Subsystem response (Fre) and capacity (Fec) factors were taken as those developed under the LLNL study [NRC, 1985a, 1985b], while assumptions regarding failure modes and consequences for the major components and their supports were drawn from the references (i.e., RCL major component supports were the only failure modes considered).

4.6.2 CE Plant

The staff evaluated a selected CE plant as a representative PWR, with CE as the NSSS vendor. The staff used the structure response factor, Frs, of Table 4-4, and took the subsystem response factor from NUREG/CR-3663 [NRC, 1985b] (reproduced here as Table 4-9).

Table 4-9 Factors of Safety: RCL Major Component Response

Response Factor	Basis for Response Factor	Median Factor
Spectral shape (Fess)	Spectral shape response factor based on smoothed, peak broadened, and enveloped response spectra	1.36
Damping factor (Fd)	SSE design value (1%) vs. best estimate (7%)	1.64
Modeling (Fem)	State-of-the-art methods complex systems	1.0
Analysis method (Fean)	Assume best estimate	1.0
Modal combination (Fmc)	Modes combined by SRSS	1.0
Directional combination (Fec)	Design based on maximum horizontal plus vertical earthquake components	1.15
Nonlinear (Fnl)	Assume nonlinear effects included in damping factor and capacity factor	1.0
Total (Fre)		2.56

For this scoping evaluation, the median capacity Scale Factor for the RCL major component supports was calculated as 2.0 [NRC, 1985b]. Consequently, the median Factor of Safety for the fragility evaluation of indirectly induced failure of RCL piping is as follows:

$$F = Frs * Fre * Fc = 2.33 * 2.56 * 2.0 = 11.93 \tag{4-12}$$

$$Ac = F * Asse = 11.93 * 0.15g = 1.79g \tag{4-13}$$

The approach to the scoping study was to integrate the fragility function over the seismic hazard curve using the mean seismic hazard curve of NUREG-1488 [Sobel, 1994]. The seismic hazard was then integrated from 0 to 1.5g, as specified for IPEEE evaluations, and the fragility function was defined by the median of 1.79g and a composite uncertainty of 0.62 (i.e., the value derived

in NUREG/CR-3663 [NRC, 1985b]). Integrating over the seismic hazard curve from 0g to 1.5g, the mean probability of failure of the lowest capacity component support was 1.65×10^{-6} per year.

4.6.3 Westinghouse Plant

The staff evaluated an indirectly induced piping failure of one Westinghouse plant, which NUREG/CR-3660 [NRC, 1985a] identified as being lower bound in terms of seismic capacity. The plant comprised two units, denoted Plants 8 and 9, which are founded on rock. Notably, the SSE design was based on NRC Regulatory Guide 1.60 design response spectra, anchored to a horizontal PGA of 0.25g. In the same manner detailed above, the staff developed the median Factor of Safety for the lower bound failure mode for the plants' major component supports, as follows:

$$F = F_{rs} * F_{re} * F_c = 2.12 * 1.16 * 3.06 = 7.52 \quad (4\text{-}14)$$

$$A_c = F * A_{sse} = 7.52 * 0.25g = 1.88g \quad (4\text{-}15)$$

To calculate the Frs parameter, the staff selected the seismic hazard curve for a plant site with rock conditions and a high design SSE value.

As for the CE plant, the staff's approach to the scoping study was to integrate the fragility function over the seismic hazard curve. Integrating over the seismic hazard curve from 0g to 1.5g, as specified for IPEEE evaluations, the staff determined that the mean probability of failure of the lowest-capacity component support is 7.6×10^{-7} per year. The staff also defined the fragility function by the median of 1.88g and a composite uncertainty of 0.42 (i.e., the value derived in NUREG/CR-3663 [NRC, 1985b]).

The staff also considered a further sensitivity study to assess the effect of a variation in composite uncertainty (i.e., a composite uncertainty of 0.62, as used for the CE plant). Integrating over the seismic hazard curve from 0g to 1.5g, the staff determined that the mean probability of failure of the lowest-capacity component support is 2.74×10^{-6} per year for this sensitivity study.

4.6.4 Observations

For the two cases considered, indirectly induced piping failure attributable to major component support failure has probabilities of occurrence of less than 10^{-5} per year — a threshold of interest. The staff adopted the assumptions of NUREG/CR-3660 and NUREG/CR-3663 [NRC, 1985a, 1985b], and the conclusions are dependent on those assumptions. Differences in methodology are attributable to the updated seismic hazard curves, updated UHS, and evaluation for a mean estimate only (i.e., no separate treatment of randomness and modeling uncertainty, and no ability to consider a family of seismic hazard curves with appropriate weights). Nonetheless, the results indicate that indirectly induced piping failure is unlikely to govern the combined failure of piping.

5 SUMMARY AND CONCLUSIONS

5.1 Summary

This report describes the approaches used and results obtained in evaluating the potential effects of seismic loading on the proposed TBS. This study used different approaches in evaluating unflawed piping, flawed piping, and indirect failures of other components and component supports that could lead to piping failure. For unflawed and flawed piping, the approach used is a hybrid deterministic and probabilistic analysis, which addresses uncertainties (in part) through sensitivity analyses, rather than explicit uncertainty analysis. For indirect failures, the approach is based on an earlier study by LLNL [NRC, 1985a, 1985b, 1985c, and 1988a].

To illustrate the effects of seismic loading on the likelihood of failure of unflawed piping, the staff evaluated several specific cases of RCS piping from selected PWRs, with diameters larger than the TBS. The choice of PWRs was based on the availability of information on design stresses and material properties. The evaluations of unflawed piping included an assessment of seismic stresses, which eliminates some of the conservatism inherent in the design process. The failure criterion used for unflawed piping was based on 3- to 6-inch (7.6- to 15.2-cm) diameter component tests, which may be conservative for this situation.

The staff also conducted flawed piping analyses for RCS piping of selected PWRs, based on the availability of design stress and material property information. Piping and weld materials included stainless steel, ferritic steel, and cast austenitic stainless steel that is moderately sensitive to thermal aging. Two different types of analyses were conducted for large-diameter PWR piping systems.

The first type was a surface flaw evaluation procedure. For this evaluation, the staff performed two sets of analyses for each flawed piping configuration. The first analysis determined the ASME Code allowable flaw size[***] using the design SSE loads for seismic loading and the calculational procedures described in Appendix C to Section XI of the ASME Code. (This ASME Code allowable surface flaw size represents the largest surface flaw that would be permitted in a pipe at the end of the evaluation period.) The second analysis estimated the critical surface flaw size for the piping, which is defined as the flaw size that would lead to a failure (i.e., complete severance) of the piping under certain stress conditions. For this analysis, the staff estimated the critical surface flaw sizes using applied stresses corresponding to normal

---

[***] Appendix C to Section XI of the ASME BPV Code provides a method for determining the acceptability of flawed piping (identified during an in-service inspection) for continued service for a specified time period or to determine the time interval until a subsequent inspection. For an Appendix C analysis, an "allowable flaw size" (depth for circumferential flaws and length for axial flaws) is determined using safety Scale Factor multipliers on the applied loading (stresses). These safety Scale Factors vary by the type of operating condition used in the analysis; normal operating conditions are referred to as Service Level A and seismic loading is denoted by Service Level D in Appendix C. This allowable flaw size is the maximum flaw extent for which continued operation is permitted under the ASME Code; operation with flaws greater than this allowable flaw size is unacceptable. This allowable flaw size is compared to the flaw size calculated for the operating period under consideration using the initial flaw configuration (from a flaw identified in an ISI or an assumed largest flaw that could reasonably be missed by the ISI) with growth of this flaw by all relevant mechanisms, such as fatigue or stress corrosion cracking.

For the analyses described in this report, the ASME Code allowable flaw size calculations use the SSE loads for the seismic loading; yield strength values from Section II of the ASME Code, and mean values from typical data for each material type.

operations and earthquake events associated with 10^{-5} and 10^{-6} annual probabilities of exceedance.

The seismic stresses were the same as those used for the unflawed piping analysis, with the exception of a correction factor for nonlinear behavior in the piping steel. Additionally, the analysis of flawed piping relaxed some of the conservatism inherent in ASME flaw evaluation based on the requirements set forth in Section XI of the ASME BPV Code. For example, the safety factors for stresses were reduced to 1.0 for these critical surface flaw size estimates. As such, these estimates are characterized as "best-estimate" and can be considered to be conservatively biased but not necessarily a lower bound. Comparison of the ASME Code allowable flaw size and best estimate flaw size determines the limiting flaw size from the TBS perspective.

The second type of evaluation was for a through-wall flaw to examine the viability of a leak detection approach for loading higher than the design-basis earthquake. The staff determined the leakage flaw size at specific leak rates associated with the normal operating conditions, and then compared this leakage flaw size to the critical through-wall flaw sizes associated with stresses corresponding to normal operations and earthquake events associated with 10^{-5} and 10^{-6} annual probabilities of exceedance. The staff used the same applied correction factors on the seismic stresses as in the surface flaw evaluations, with different safety factors on the leakage crack length. Initially, analyses were conducted with a safety factor of 10 on the technical specification leak rate (1 gpm for PWRs) with different types of cracking mechanisms (i.e., air fatigue cracks, corrosion fatigue cracks, and PWSCC cracks). The staff selected five PWR piping systems from the 52 piping systems analyzed for the surface flaw evaluations to represent different ratios of normal to normal+10^{-5} seismic stresses. The ratio of the normal to normal+10^{-5} seismic stresses is a significant factor in this evaluation, and the cases chosen effectively covered the total range of that stress ratio for the 52 piping cases. Subsequently, the staff conducted analyses with a safety factor of 1.5 on the crack length for the 10^{-5} seismic stresses and a safety factor of 1.0 for the 10^{-6} seismic stresses. In addition, the staff conducted a sensitivity analysis using a leakage rate of 0.5 gpm.

The approach used for indirect failures is a scoping evaluation based on the earlier LLNL study [NRC, 1985a, 1985b, 1985c, and 1988a]. The primary reason for limiting the scope of this approach is that there is no fundamental change required in the approach used in the LLNL study. However, evaluations are needed to assess whether different seismic hazard information affects the earlier results and conclusions. In this approach, components, component supports, and piping supports are selected based on whether their failure could lead to a piping failure. Seismic fragilities (conditional probabilities of failure given a seismic event of certain intensity) are estimated for these items, and then these fragilities are convolved with the seismic hazard (frequency of exceeding seismic events of a given size per year — also termed annual probability of exceedance) to obtain a distribution of failure probability. Two component support cases were selected from the earlier LLNL study to see how their fragilities and failure probabilities will be affected by using different hazard information.

In all of the above analytical activities, the staff used mean seismic hazard curves from the LLNL hazard estimates developed in the early 1990s [Sobel, 1994]. This report discusses the reasons for using these hazard curves.

In addition, the staff reviewed past earthquake experience reports, associated data, and past seismic PRAs to glean additional insights. The staff then used those insights to evaluate results of the analytical investigations performed in this study.

In order to evaluate the results and conclusions of this study in their proper context, it is important to recall how the TBS was determined. As discussed in Section 1.2, the TBS was determined by considering the results of an expert elicitation to estimate the frequency of different size pipe breaks owing to service degradation of piping subjected to normal operating loads [NRC, 2005]. As a starting point in determining the TBS, the staff used the break size related to a nominal failure frequency of 10^{-5} per year. However, to address uncertainties in the elicitation, the staff adjusted (i.e., increased) this size, resulting in a TBS that does not directly relate to a specific break frequency, but is expected to have a frequency not greater than 10^{-5} per year.

The present study used an earthquake frequency of 10^{-5} per year as a general point of comparison to the TBS. However, the staff also considered earthquakes of a lower frequency and larger size to assess the sensitivity of pipe failure frequency to earthquake size and to demonstrate acceptable margins even for those less-likely earthquakes.

The following section presents key conclusions and an overall assessment for unflawed piping, flawed piping, and indirect failures.

5.2   **Conclusions and Overall Assessment**

<u>Unflawed Piping</u>: Analyses performed in this study show that failure probabilities of unflawed piping, as defined in Section 2.1, are significantly low compared to the frequency of 10^{-5} per year used as a basis to establish the TBS.

<u>Flawed Piping</u>: The following are some of the higher-level observations, conclusions, and insights derived from results discussed in Section 4.5:

(1) The absolute size of the best-estimate critical surface flaw sizes are large for ground motion levels corresponding to seismic events associated with 10^{-5} and 10^{-6} annual probabilities of exceedance. For the cases analyzed, even for very long circumferential cracks, the surface flaw depth must be larger than 40% of the wall thickness in order to become critical for stresses associated with a 10^{-5} annual probability of exceedance seismic event. The corresponding surface flaw depth is 30% of the wall thickness for a seismic event with a 10^{-6} annual probability of exceedance. This is the most significant finding, in that these surface flaw sizes are large enough that either an effective inspection program or a leak detection system can be implemented to ensure that a surface flaw will be detected in time and will not grow to critical size during service. The probability of having such flaws in service is mitigated by considering the following factors:

- Inservice inspections are generally capable of detecting much smaller flaws.
- Generally, the past practices have been to repair a flaw on discovery for most materials used in primary loop piping, or to mitigate conditions that lead to flaw growth.
- In general, analyses conducted using an approach similar to that used for LBB evaluations have demonstrated that leakage flaw sizes are much smaller than critical flaw sizes, and detectable leaks can occur before flaws reach critical size. At high stresses associated with a 10^{-5} annual probability of exceedance seismic event, these lengths are on the order of about 15% of the pipe circumference, while in LBB analyses, typical leakage flaw sizes in PWR primary piping are about 6% of the circumference. Hence, for degradation mechanisms where long surface cracks

do not develop, some protection would naturally occur from leakage detection. However, leakage protection may not necessarily occur for flaws developed from stress-corrosion cracking mechanisms where long surface cracks could develop.

(2) Critical flaw sizes depend on applied stresses and material strength parameters and vary significantly for different locations and piping systems. Both normal and seismic stresses depend on layout configuration, design process, support designs, site seismic hazard and ground motion characteristics, and other plant-specific features.

(3) The flaw sizes allowed by the ASME Code (see footnote*** in Section 5.1) are affected by whether one uses ASME Section II material properties or actual material properties based on available test data. (Appendix C to Section XI of the ASME Code permits the use of actual material properties.)

(4) Comparisons of best-estimate versus Code-allowable flaw size curves for analyzed cases fall into three categories: (a) best-estimate flaw sizes are bigger than the maximum allowable flaw sizes using either Code material properties or actual material properties; (b) best-estimate flaw sizes fall between two Code flaw size estimates; and (c) best-estimate critical flaw sizes are smaller than both Code estimates. Thus, the best-estimate flaw sizes are not necessarily larger than the Code flaw sizes in all cases; however, more frequently, the maximum flaw size allowed by the Code was smaller than the best-estimate critical flaw sizes.

(5) Piping systems of BWR and west coast plants have not been studied in this report because required information was not readily available. However, there are no inherent limitations in applying the approach used here to piping systems of BWR and west coast plants.

Indirect Failures: The present study yielded the following key observations and conclusions:

(1) As with other parts of the seismic analysis, indirect failure evaluations are plant and site-specific. Details of plant layout and component and pipe support designs vary significantly from plant to plant.

(2) For the two cases analyzed, indirectly induced piping failure attributable to major component support failure has a mean failure probability on the order of 10^{-6} per year.

In summary, this study has demonstrated that the critical flaws associated with the stresses induced by seismic events of 10^{-5} and 10^{-6} annual probability of exceedance are large, and coupled with other mitigative aspects, the probabilities of pipe breaks larger than the TBS are likely to be less than 10^{-5} per year. Similarly, for the cases studied, the probabilities of indirect failures of large RCS piping systems are less than 10^{-5} per year.

The intent of the staff study was not to perform bounding analyses that will encompass all potential variations, including site-to-site variability in the seismic hazard. The purpose of the staff study was to get a measure of seismic effects on the proposed TBS and to provide information on key considerations to facilitate the public review and comment period to elicit comments and information germane to this issue.

As stated on page 67618 of the *Federal Register* notice announcing the proposed rule on Risk-Informed Changes to Loss-of-Coolant Accident Technical Requirements (70 FR 67598; November 7, 2005), the NRC requests specific public comments on the results of the evaluation contained in this report. The NRC also requests specific public comments on the effects of pipe degradation on seismically induced LOCA frequencies and the potential for affecting the selection of the TBS and on any other potential options and approaches to address this issue. The staff will use this information in finalizing the technical basis as it moves forward with the rulemaking.

6   REFERENCES

ANS, "American National Standard – External Events PRA Methodology," ANSI/ANS 58-21-2003, American Nuclear Society, La Grange Park, IL, March 2003.

Brust, F.W., et al., "Assessment of Short Through-Wall Circumferential Cracks in Pipes," NUREG/CR-6235, U.S. Nuclear Regulatory Commission, Washington, DC, April 1995.

EERI, "Kocaeli, Turkey, Earthquake of August 17, 1999 Reconnaissance Report," Supplement A to Volume 16, Earthquake Spectra, EERI No. 2000-03, Earthquake Engineering Research Institute, Oakland, CA, December 2000.

EERI, "Chi-Chi, Taiwan, Earthquake of September 21, 1999 Reconnaissance Report," Supplement A to Volume 17, Earthquake Spectra, EERI No. 2001-2, Earthquake Engineering Research Institute, Oakland, CA, April 2001.

EERI, "Southern Peru Earthquake of 23 June 2001 Reconnaissance Report," Supplement A to Volume 19, Earthquake Spectra, EERI No. 2003-1, Earthquake Engineering Research Institute, Oakland, CA, January 2003.

EPRI, "Evaluation of Flaws in Austenitic Steel Piping," (Technical basis document for ASME IWB-3640 analysis procedure), prepared by Section XI Task Group for Piping Flaw Evaluation, EPRI Report NP-4690-SR, Electric Power Research Institute, Palo Alto, CA, April 1986.

EPRI, "Evaluation of Flaws in Ferritic Piping," EPRI Report NP-6045, prepared by Novetech Corporation for the Electric Power Research Institute, Palo Alto, CA, October 1988.

EPRI, "A Methodology for Assessment of Nuclear Power Plant Seismic Margin," EPRI Report NP-6041-SL, Rev. 1, Electric Power Research Institute, Palo Alto, CA, 1991.

EPRI, "Piping and Fitting Dynamic Reliability Program - Program Summary," EPRI TR-102792, Vols. 1–5, Electric Power Research Institute, Palo Alto, CA, 1995.

EPRI, "Individual Plant Examination of External Events (IPEEE) Seismic Insights," Revision to EPRI Report TR-112932, Electric Power Research Institute, Palo Alto, CA, 2000.

Ghadiali, N., "NRCPIPE, Windows Version 3.0, User's Guide," Computer Code and User's Manual for International Piping Integrity Research Group Program Members, Battelle, NRC Contract NRC-04-91-063, NRC Job Code D2060, U.S. Nuclear Regulatory Commission, Washington, DC, April 30, 1996a.

Ghadiali, N., et al., "Deterministic and Probabilistic Evaluations for Uncertainty in Pipe Fracture Parameters In Leak-Before-Break and In-Service Flaw Evaluations," NUREG/CR-6443, U.S. Nuclear Regulatory Commission, Washington, DC, June 1996b.

Ghadiali, N., and Wilkowski, G.M., "Fracture Mechanics Database for Nuclear Piping Materials (PIFRAC)," in *Fatigue and Fracture*, Volume 2, PVP-Vol. 324, pp 77–84, American Society of Mechanical Engineers, New York, NY, July 1996c.

Ghadiali, N., et al., "S Q U I R T (Seepage Quantification of Upsets In Reactor Tubes) User's Manual, Windows Version 1.1, by Battelle-Columbus for the U.S. Nuclear Regulatory Commission, Washington, DC, March 24, 2003.

Jaquay, K., "Seismic Analysis of Piping," NUREG/CR-5361, U.S. Nuclear Regulatory Commission, Washington, DC, June 1998.

Johnson, S., and Maslenikov, O., "Soil Structure Interaction Response of a Typical Shear Wall Structure," UCID-20122 Vols. 1 and 2, Lawrence Livermore National Laboratory, Livermore, CA, 1984.

Kumar, V. and German, M., "Elastic-Plastic Fracture Analysis of Through-Wall and Surface Flaws in Cylinders," EPRI Report NP-5596, Palo Alto, CA, January 1986.

Kupperman, D.S., et al., "Barrier Integrity Research Program: Final Report," NUREG/CR-6861, U.S. Nuclear Regulatory Commission, Washington, DC, December 2004.

Marschall, C.W., and Wilkowski, G.M., "Effect of Cyclic Loads on Ductile Fracture Resistance," in ASME Special Technical Publication, Vol. 166, pp. 1-14, American Society of Mechanical Engineers, New York, NY, July 1989.

Maslenikov, et al., "Comparison of Design and Probabilistic Analyses of Nuclear Power Plants," in *Proceedings of the 13th International Conference on Structural Mechanics in Reactor Technology*, SMiRT 13, Porto Alegre, Brazil, August 1995.

Newmark, N.M., and Hall, W., "Development of Criteria for Seismic Review of Selected Nuclear Power Plants," NUREG/CR-0098, U.S. Nuclear Regulatory Commission, Washington, DC, 1978.

Norris, D., and others, "PICEP: Pipe Crack Evaluation Program," EPRI Report NP-3596-SR, Palo Alto, CA, 1984.

NRC, Regulatory Guide 1.61, "Damping Values for Seismic Design of Nuclear Power Plants," U.S. Nuclear Regulatory Commission, Washington, DC, October 1973.

NRC, "Probability of Pipe Failure in the Reactor Coolant Loops of Westinghouse PWR Plants," NUREG/CR-3660, UCID-19988, Vols. 1–4, U.S. Nuclear Regulatory Commission, Washington, DC, 1985a.

NRC, "Probability of Pipe Failure in the Reactor Coolant Loops of Combustion Engineering PWR Plants," NUREG/CR-3663, UCRL-53500, Vols. 1–3, U.S. Nuclear Regulatory Commission, Washington, DC, 1985b.

NRC, "Probability of Pipe Failure in the Reactor Coolant Loops of Babcock and Wilcox PWR Plants," NUREG/CR-4290, UCRL-53644, Vols. 1–3, U.S. Nuclear Regulatory Commission, Washington, DC, 1985c.

NRC, "Probability of Failure in BWR Reactor Coolant Piping," NUREG/CR-4792, UCID-20914, Vols. 1–4, U.S. Nuclear Regulatory Commission, Washington, DC, 1988a.

NRC, Generic Letter 88-20, Supplement 4, U.S. Nuclear Regulatory Commission, Washington, DC, 1988b.

NRC, "Perspectives Gained from the Individual Plant Examination of External Events (IPEEE) Program," NUREG-1742, Vols. 1 & 2, U.S. Nuclear Regulatory Commission, Washington, DC, 2001.

NRC, "Estimating Loss-of Coolant Accident (LOCA) Frequencies through the Elicitation Process," Draft NUREG-1829 for public comment, U.S. Nuclear Regulatory Commission, Washington, DC, June 2005.

Paris, P.C., and Tada, H., "Application of Fracture Proof Methods Using Tearing Instability Theory to Nuclear Piping Postulating Circumferential Through Wall Cracks," NUREG/CR-3464, U.S. Nuclear Regulatory Commission, Washington, DC, September 1983.

Paul, D.D., et al., "Evaluation and Refinement of Leak-Rate Estimation Models," NUREG/CR-5128, Rev. 1, U.S. Nuclear Regulatory Commission, Washington, DC, June 1994.

Rahman, S., Ghadiali, N., Paul, D., and Wilkowski, G., "Probabilistic Pipe Fracture Evaluations for Leak-Rate-Detection Applications," Topical Report, NUREG/CR-6004, U.S. Nuclear Regulatory Commission, Washington, DC, April 1995a.

Rahman, S., Brust, F., Ghadiali, N., Choi, Y.H., Krishnaswamy, P., Moberg, F., Brickstad, B., and Wilkowski, G., "Refinement and Evaluation of Crack-Opening Analyses for Short Circumferential Through-Wall Cracks in Pipes," NUREG/CR-6300, U.S. Nuclear Regulatory Commission, Washington, DC, April 1995b.

Reed, J., and Kennedy, R.P., "Methodology for Developing Seismic Fragilities," Prepared for Electric Power Research Institute, EPRI TR-103959, Palo Alto, CA, June 1994.

Rudland, D., et al., "The Effects of Cyclic and Dynamic Loading on the Fracture Resistance of Nuclear Piping Steels," NUREG/CR-6440, U.S. Nuclear Regulatory Commission, Washington, DC, December 1996.

Rudland, D., et al., "Comparison Of Estimation Schemes and FEM Analysis Predictions Of Crack-Opening Displacement for LBB Applications," PVP Volume 410-1, pp. 157–162, American Society of Mechanical Engineers, New York, NY, July 2000.

Scott, P.M., et al., "Fracture Evaluations of Fusion-Line Cracks in Nuclear Pipe Bimetallic Welds," NUREG/CR-6297, U.S. Nuclear Regulatory Commission, Washington, DC, April 1995.

Scott, P.M., "Development of Technical Basis for Leak-Before-Break Evaluation Procedures," NUREG/CR-6765, U.S. Nuclear Regulatory Commission, Washington, DC, May 2002.

Scott, P.M., et al., "Technical Development of Loss of Coolant Accident Frequency Distribution Program - Subtask 1a: Finalize and QA SQUIRT Code," Battelle-Columbus report to NRC-RES on NRC Contract No. RES-02-074, December 2003.

Scott, P.M. et al., "The Battelle Integrity of Nuclear Piping (BINP) Program Final Report," NUREG/CR-6837, Vol. 2, U.S. Nuclear Regulatory Commission, Washington, DC, June 2005.

Sezen, H., et al., "Structural Engineering Reconnaissance of the August 17, 1999 Earthquake: Kocaeli (Izmit), Turkey," Pacific Earthquake Engineering Research Center, College of Engineering, University of California, Berkeley, CA, PEER 2000/09, December 2000.

Slagis, G., "Review of Seismic Response Data for Piping," Pressure Vessel Research Committee, American Society of Mechanical Engineers, New York, NY, June 1995.

Sobel, P., "Revised Livermore Seismic Hazard Estimates for Sixty-Nine Nuclear Power Plant Sites East of the Rocky Mountains," NUREG-1488, .S. Nuclear Regulatory Commission, Washington, DC, April 1994.

Stevenson, J.D., "Addendum, Summary, and Evaluation of Historical Strong-Motion Earthquake Seismic Response and Damage to Above-Ground Industrial Piping," NUREG-1061, Vol. 2, U.S. Nuclear Regulatory Commission, Washington, DC, February 1985.

Stevenson, J.D., "Survey of Strong Motion Earthquake Effects on Thermal Power Plants in California with Emphasis on Piping Systems," Main Report, Vol. 1, NUREG/CR-6239, ORNL/Sub/94-SD427/2/V1, Stevenson and Associates for Oak Ridge National Laboratory, U.S. Nuclear Regulatory Commission, Washington, DC, November 1995.

Wilkowski, G.M., et al., "Analysis of Experiments on Stainless Steel Flux Welds," NUREG/CR-4878, U.S. Nuclear Regulatory Commission, Washington, DC, April 1987.

Wilkowski, G.M., et al., "Degraded Piping Program - Phase II, Summary of Technical Results and Their Significance to Leak-Before-Break and In-Service Flaw Acceptance Criteria," NUREG/CR-4082, Vol. 8, U.S. Nuclear Regulatory Commission, Washington, DC, March 1989.

Wilkowski, G.M., et al., "Proposed Modification of the ASME Section XI Pipe Flaw Evaluation Criteria Based on New Surface-Cracked," *Fatigue and Fracture*, Volume 1, pp. 51–64, American Society of Mechanical Engineers, New York, NY, July 1996.

Wilkowski, G.M., et al., "International Piping Integrity Research Group (IPIRG) Program, Final Report," NUREG/CR-6233, Vol. 4, U.S. Nuclear Regulatory Commission, Washington, DC, June 1997.

Wilkowski, G.M., et al., "State-of-the-Art Report on Piping Fracture Mechanics," NUREG/CR-6540, BMI-2196, U.S. Nuclear Regulatory Commission, Washington, DC, February 1998.

Wilkowski, G.M., et al., "Impact Of PWSCC And Current Leak Detection On Leak-Before-Break Acceptance," Paper PVP2005-71200, *Proceedings of ASME-PVP 2005 ASME/JSME Pressure Vessels And Piping Conference,* Denver, CO, July 17–21, 2005.

Williams, C., et al., "The Impact of Fracture Toughness and Weld Residual Stresses of Inconel 82/182 Bimetal Welds on Leak-Before-Break Behavior," *Proceedings of ASME-PVP 2004, ASME/JSME Pressure Vessels And Piping Conference*, La Jolla, San Diego, CA, July 2004.

www.ingramcontent.com/pod-product-compliance
Lightning Source LLC
Chambersburg PA
CBHW081140170526
45165CB00008B/2741